TO SPEAK
FOR THE TREES

TO SPEAK FOR THE
TREES

My Life's Journey from Ancient Celtic Wisdom to a Healing Vision of the Forest

DIANA BERESFORD-KROEGER

Timber Press | Portland, Oregon

Paperback edition published in the United States, the United
Kingdom, Ireland, New Zealand, Australia, and South Africa in 2021
by Timber Press, Inc.

The Haseltine Building
133 S.W. Second Avenue, Suite 450
Portland, Oregon 97204-3527
timberpress.com

Text and cover design by Adrianna Sutton
Text is set in Capita, a typeface designed by Dieter Hofrichter in 2013

ISBN 978-1-64326-132-4

Printed in the United States of America
Catalog records for this book are available from the
Library of Congress and the British Library.

MIX
Paper from
responsible sources
FSC
www.fsc.org   FSC® C008955

———

To my ancestors at the Castle of Ross, Killarney,

who lived in Lackavane and the Valley of Lisheens.

You gave me my greatest gift, that of the mind.

———

CONTENTS

Brehon

The countryside of Éire
has been put to sleep with poetry, long ago.
Fadó.
Fadó.
The land is murmuring.
Soft fields of dreams move slowly
filled with hares and long tail grass,
mountains strained with purple heathers
and foxes fevered with glints of yellowed furze.
Woodland words,
ragged with rain,
between the corncrake sky
and the fishtail sounds
sucking up the sea.
The yoke of the moon will shield the Laws of Liberty,
arís agus arís,
pointing us straight to
Brehon.

DIANA BERESFORD-KROEGER

INTRODUCTION

I HAVE ALWAYS found it difficult to think about the story of my life, let alone tell it. I suffered great traumas as a child. To protect myself, I took my pain and put it down a deep well in my mind. I hid it from myself so that I could function, and I moved through my entire scientific education and decades of research with my eyes always cast ahead, looking for the next question, the next answer, the next piece of understanding and wisdom.

But the person I am today could not exist without that trauma. It led me, as a thirteen-year-old girl, on stepping stones to one of the last bastions of the Celtic culture in Ireland, a place called the Lisheens Valley in County Cork. I arrived in Lisheens in need of something to help hold me together just as the place itself was falling apart. The ancient knowledge of the Druids and the Brehon Laws, kept safe, refined and handed down from one generation to the next for millennia, was on the verge of being lost. Instead, it was given to me, an understanding of the healing powers of plants and the sacred nature of the natural world that remains the greatest gift I have ever received.

The only thing asked of me in exchange for that gift was that I not keep it to myself. And though I have shared my ideas and discoveries freely during my fifty-year career in science, I have always held pieces of my story back, keeping the complete picture obscured even from myself.

But now we find ourselves in a special time. On the one hand, climate change poses the most significant threat to our planet that humanity has ever faced. On the other, we are better equipped than ever before to take on that challenge. To do so, though, we need to understand the natural world as people once did. We need to see all that the sacred cathedral of the forest offers us, and understand that among those offerings is a way to save our world.

We are all woodland people. Like trees, we hold a genetic memory of the past because trees are parents to the child deep within us. We feel that shared history come alive every time we step into the forest, where the majesty of nature calls to us in a voice beyond our imaginations. But even in those of us who haven't encountered trees in months or even years, the connection to the natural world is there, waiting to be remembered.

In telling the story of my life and the leaves, roots, trunks, bark, and stems that weave all through it, I hope to stir that memory. I want to remind you that the forest is far more than a source of timber. It is our collective medicine cabinet. It is our lungs. It is the regulatory system for our climate and our oceans. It is the mantle of our planet. It is the health and wellbeing of our children and grandchildren. It is our sacred home. It is our salvation.

Trees offer us the solution to nearly every problem facing humanity today, from defending against drug resistance to halting global temperature rise, and they are eager to share those answers. They do so even when we can't or won't hear them. We once knew how to listen. It is a skill we must remember.

PART I

Comfort in a Stone

MY WEEPING STONE sat on the highest shoulder of the valley, where it pointed to the blue above. The stone was way taller than my head, a huge rectangle except for the curve at its crest where chunks had broken loose long ago. Its surface was weathered into rough ripples interrupted by the rounded scabs of lichens. The stone was easily twice the size of the heavy dealwood table in the farmhouse kitchen, big enough that any changes to it occurred on a timeline far too slow for me to perceive, which gave it a welcome constancy.

I called it my weeping stone because I trudged up the hill to be by its side when I felt especially alone. I never really cried. I was beyond tears. Or I repressed my tears, never noticing because I swallowed them whole. I would sit at the stone's base and lean back against its sturdy flank, ready to slip around to another side and hide if anyone from below called out to me—a reassuring defence, even if no one ever did call.

As I sat there, the slow throb of the Earth settled its calm into my bones. Below me was the farmhouse with its puffs of smoke and, beyond it, the fields of my great-aunt's farm, each one named in Gaelic like an ancient song. Our neighbours' farms blanketed both sides of the valley in a patchwork that glowed with a green that seemed to have fire in it.

I could watch the seabirds spread open the timothy veil of the pastures and sometimes see the Owvane River, packed with salmon, at the heart of the valley spilling westwards to the open arms of Bantry Bay. If I turned north, I could admire the great sleeping silhouettes of the Caha mountains, colours dancing on their hulking forms. *Cnoc Buí*—the yellow hill—electrified by its yellow flowers, seemed to vibrate with the chrome of gorse. At times, as I watched the aquamarine of the sea, I wondered about the bolts of bronze that came and went in a silent symphony of colour. From that vantage point, I could in fact see the entire landscape that had sustained my mother's family in body and in soul for the past three thousand years. The light playing with the clouds, the salt wind and the rain soothed me. While I never cried buckets up there against my weeping stone, I was a child with no shortage of pain.

On this particular summer day I'm remembering, I had climbed to the stone carrying thoughts of my father. I was an orphan, having recently lost both my parents. I had been lonely most of my young life— separated from most of those around me by nationality, religion and class, just for starters—and I had learned to live with that isolation. But my parents' deaths struck a blow that I wasn't sure I could recover from. Months and months had passed and still I felt numb. The daily freshness of their deaths was disorienting, as though the ground and sky had been pulled out from under me. My mourning for my father was constant, the loss so strong that at times I felt winded by its power over me. Some vital part of me was missing and would never come back, because death had closed a door. I just wanted to be small, only a dot, a tiny one. Maybe if I held my breath I could disappear altogether.

I huddled into myself at the base of the stone to survive. The sight of the valley below made me feel both safe and like a tiny dot, as small as the black and white cows down there, moving slowly with their pink udders swaying. They were content. I must be, too. And, as I calmed down, I was able to take sober stock of my life.

On my father's side, I was a descendant of the English aristocracy, the most fragile leaf on a Beresford family tree that included earls, lords and marquises by the branchful. On my mother's side, I was as Irish as the heather in front of me, the last living drop of a bloodline that could be traced back to the kings of Munster. My dual heritage had inspired resentments, the consequences of which I have borne all my life. As a female child among the Beresfords, I faced the stumbling block of primogeniture. I could not inherit anything of value from my father's estate other than my bloodline and my name. I was a crossbreed, too Irish for the English and too English for the Irish. My one saving grace in Irish eyes was that I was a female and therefore more important than the male.

Thoughts that my father's family would continue to ignore me, as they had since my father's death, sent me into a panic. But that passed, too, as I looked across the pastures of the Valley of Lisheens, the handful of square miles of rural Ireland where I would spend my summers for the next decade. I had no inkling yet of the hope that existed right in front of me or of the ways the land and its people would guide and shape me. I didn't know that the older generation of my mother's family had already met down at Pearson's Bridge to discuss my fate. I didn't know that they had already decided to give me the gift of their ancient knowledge, their open secret, and that it would save my life. Or that they intended me to become their "child of destiny." All I knew, leaning against my weeping stone, was that I was invisible, crushed from too many deaths and utterly alone.

———

My parents, Eileen O'Donoghue and John Lisle de la Poer Beresford, met in England, most probably in London sometime during the Second World War, and fell in love. Though, as a child, I quite enjoyed digging up the romantic past of all those around me, I never got a chance to ask my parents for their love story. A few tidbits I did know. One was that my mother, in an evening gown with silver silken gloves to her elbows, adorned with seed pearls and sapphires, was hard to resist. Once, when I was young, I was brought in to watch her sweep across the dance floor at a private ball. Everyone present gave way to her elegance and beauty. Why a man might fall for a woman such as my mother isn't hard to parse in its broad strokes.

Jack, my father, came from the best of everything. He was an Eton boy who had been presented at court, the son of Lord William Beresford; he was related to the Churchills and Spencers and all the rest of English high society. He lived the ultimate aspirational lifestyle of the early twentieth century, and that alone would have drawn women's interest, but he was also a charming, cultured man. Even my mother's people in Lisheens spoke of him with a grudging admiration and fondness, despite his status among the Protestant Anglo-Irish elite. He was a linguist, fluent in thirteen languages including three dialects of Arabic, who taught at Cambridge. He was tall and he wore a monocle, which sounds silly now but suited him, I think.

Though my mother's pale skin gave her face a soft, delicate quality, she was spirited and adventurous, well read and outgoing—capable of commanding a room when she felt like it. And she was athletic, an accomplished horsewoman who'd ridden to school every day of her childhood. She had a wild streak and an uncommon bond with animals, especially horses and their kin, both qualities captured in my favourite story about her, which was that when she was a girl, she once managed to get a donkey onto the roof of her schoolhouse. No one could figure out how she did it and, so the story goes, she never told.

Neither of her parents lived to see her marry an English aristocrat, but like her surviving relatives, they would have seen it as a divisive act of disobedience. My father's family, in contrast, preferred judgment of the silent type.

Though we later lived in Bedford, England, for the earliest years of my life, I was born in Islington, a part of greater London, in the summer of 1944. My first memory is of being breastfed. I remember my mother's nipple touching the top of my palate and instantly going into ecstasy, then dropping off to sleep. I may have held onto that moment for so long because of the simple pleasure and contentedness I felt. More likely, though, it has remained with me because times of true connection with my mother were so rare.

When I was two or three, my parents began to make regular trips to Ireland, towing me along like luggage. Travelling the world and summering in the countryside was merely what people of their class did. Regardless of their taboo cross-cultural union and my mother's rebellious streak, on the whole my parents did what was expected. There was one point, however, on which my mother refused to bend: she insisted, against the proper English wishes of my proper English father, on taking me year after year to visit our ancestral seat in Ireland.

The two of us would travel by car to the border between the counties of Kerry and Cork, where we'd slow to a reverential pace before taking the narrow road up to the Pass of Keimaneigh. At the pinnacle, where the pass made its definitive cut through the rock, the mountains almost touched above us. There, my mother would stop and we'd get out of the car, looking up in awe at the boulders that seemed held in place by just a string of heather, flushed deep purple as if straining from the effort. Over the sound of the twin streams of water rushing across the black rock on either side of the pass, my mother would tell me the legend of the priest who'd used the pass to make a daring escape during the time of the Penal Laws. Those laws, inflicted on the Irish by their English occupiers for five hundred years, until 1916, made it illegal, among other things, for any "person of the popish religion" to run a school or teach children. The priest had not only educated the local kids, he'd done it out in the open in what the locals called a "hedge" school, with scouts on watch for trouble. Threatened with imprisonment or worse, and being chased by mounted English soldiers and their dogs, he had leapt the gap at the top of the pass and successfully escaped.

After a stop at Gougane Barra to meditate in the caves that the monks had inherited, we quietly entered the gentle space of St. Finbarr's Oratory, which stood on its own sacred island. Then we swept northwest into Kerry and on to the family seat at the Castle of Ross on the shore of Loch Léin, the largest of the three lakes of Killarney. Rising from the car, my mother would light a cigarette, a Woodbine, and inhale, then release a tendril of smoke as she bent to smooth the skirt of her Parisian suit. Stepping carefully past the mud puddles, she'd look the building over with an appraising eye, as though she was a prospective buyer. She'd take in the top of the castle, open to the elements since the roof had been removed in penal times to reduce the rent, and the sow lying against one stone wall with her litter of squealing pink piglets. Stubbing out her cigarette and turning back to the car, she'd fire a parting shot: "Nobody has fixed the roof yet."

These pilgrimages were proof that my mother couldn't completely let go of her Irish heritage—that she still felt the pull of the ancient places, as well as some responsibility to allow me a connection to our past. But almost everywhere else, she dismissed the culture and beliefs of her parents as backward and full of superstition. She expected me to grow into a woman who was attractive and acceptable to my father's people, and then to make a good marriage. Otherwise, she expected me to stay silent and out of the way. So that's what I did, to the best of my ability.

When I was seven, my parents had a huge fight and separated. My father stayed in England, while my mother and I moved into a tall Georgian house at No. 5 Belgrave Place in Cork, Ireland. No explanation for the change or my father's sudden absence was offered to me; we were simply removed from him. This lack of communication wasn't unique to my parents. In that time and place, and particularly in that stratum of society, children were appendages who were not owed any emotional consideration. But my father's sudden disappearance from my life was a deep wound. He was a reserved man and never told me directly that he loved me, but in his quiet way he made me feel loved. He would draw me and paint me. (His oil portrait of me still sits in my living room.) I have memories of him playing the piano when I was very young. He would pause in his playing, warmly call me to the bench and lift me onto his lap. He would then lay my hands on top of his. My hands were too small to follow the movements of his fingers, but he wanted me to feel the rhythm of the music as he played. I also remember him perching me on his shoes—a foot on a foot—and dancing with me at our house in Bedford.

At Belgrave Place we lived with two of my mother's siblings. My uncle Patrick had been a famous athlete, known across Ireland as the distance runner and hurler Rocky Donoghue. A lifelong bachelor, he worked as a chemist at the city's gasworks. My aunt Biddy had suffered a broken back in early childhood and the injury had left her an invalid. She was often in the hospital, maybe three times a year, and had difficulty walking. Biddy was kind to me. She spoke warm words and took an interest in me. I grew to love her fiercely and took care of her to the best of my ability. I remember I read her all of *Jane Eyre*, over and over again. Uncle Pat seemed indifferent—not cruel or cold, even occasionally up for a chat, but focused on his own business and not much fussed by the thoughts or needs of a child or, for that matter, of anyone else in the house. Now that she was my sole caregiver, my mother made her feelings plainest of all. "You're just a nuisance," she would tell me. "And my life would be better without you."

I had few friends outside the house. My last name marked me as not only different, but also potentially dangerous. The Beresfords were among the most powerful families in Ireland. If a child in the neighbourhood or on the school grounds were to hurt, insult or accidentally run afoul of me, I might report the incident to a relation capable of ruining that child's entire family. Were they to parrot one of their parents' political views in earshot of me, there was no guarantee it wouldn't find its way to the Beresford family. For the most part, the people of Cork simply left me alone.

The house at Belgrave Place was part of a collection of ten, built as a unit in front of a large shared courtyard. The houses had been erected, likely in the 1700s, as English officers' quarters, and long before we took residence, someone had planted a small arboretum in the courtyard. This became my playground and I suppose, because I had no other companions, the trees seemed to welcome me. They became my friends. I would place my most precious doll, the one from America with the curly red wig and the porcelain face with blinking blue eyes, into the safekeeping of the giant bay tree. This was the tree in whose bole I played house, with the smell of bay leaves rich around my toy oven and all of my lesser dolls (the pecking order firmly established and ending with the limp cloth ones). Like my weeping stone, which I hadn't yet discovered, the trees comforted me with their immense size. Their presence had a dependability that felt like benevolence to me, and they were forever changing in ways I was hungry to understand. The trees kept coming into my dreams at night, too, with their long swags of shadow changing the landscape of the bedroom wall.

Two doors up from us, at No. 7, lived a man I was convinced could help me get to know the trees. Dr. Barrett was a naturopath who wore steel wire-rimmed spectacles and had no children. He lived with his wife and sister, both of whom also wore steel-rimmed spectacles, a commonality of no small import to my young mind. Many days, I would position myself in a laurel shrub across from his door, hidden from view by its curtain of spotted foliage, and wait for Dr. Barrett to arrive home. When he did I would meet him at his front door with my first prepared question and our lesson would unfold naturally from that starting point.

In the fall of that first year after my parents' row, a very curious thing happened. An immensely tall and immensely thin tree I'd had my eye on broke out all over in tiny ovoid red fruit I thought of as apples, having no other word for it. I had never seen anything like this tree before, over thirty feet tall and bursting with colourful bounty, and the belief that it must be a rare and special being occupied my mind completely. The tree spoke to me with its unusualness, and I was desperate to hear what it had to say. So I took my spot in the shrub and when Dr. Barrett stood on his doorstep to review his world, I approached and raised one of the apples towards his steel-rimmed eyes. "Are these apples safe to eat?" I asked. He told me yes, I could eat it, then explained that the treasure I held in my hand was actually a haw, the fruit of a species of hawthorn that he knew as the American haw tree, *Crataegus douglasii*. I took a bite and was met with a sweet and tangy flavour, a delicious discovery all my own.

From that point on, the arboretum became not just a place of fellowship and play, but also one of experiments and revelations. I remember Dr. Barrett telling me that the leaves of another hawthorn—the common variety I'd come to know by its Latin name, *Crataegus monogyna*—were both edible and good for your health. Armed with this information, I climbed as far as I could up the tree, enduring its thorns, and sampled its leaves for myself. They tasted like salad.

On another day, I was circling the bay laurel and stood on one of its small black seeds. The outer coat of the seed, its testa, split slightly under my foot and the fragrance it released was incredible. I picked up the seed and, with my fingernail, peeled the testa back to reveal something white and gleaming underneath. The smell intensified. It was the exact smell of the tree itself, only concentrated. I couldn't believe that the smell of the tree was contained so powerfully inside the seed. The connected wonder of that is still so clear in my head—both the feeling of discovering the link between the seed and its parent tree and the awe of the link itself.

At my insistence, Dr. Barrett also taught me the Latin names of any species we encountered together. This information was reward enough in itself, but every time I committed something new to memory, Dr. Barrett would produce a brown bag of chocolate-covered dates and invite me to take one. He was a very kind man.

Later that fall I began school, where I was taught by a big-boned, red-faced woman named Miss Barrett, the head of the school. She wasn't related to the naturopath, but the positive association I had formed with the name made her feel familiar and safe to me. When Miss Barrett asked me about my summer holidays, with a precocious flourish, I recited for her the Latin names of every tree visible out the classroom window. I don't remember her reaction in the moment, but I do know that she sent my mother a letter.

On the Saturday afternoon of that same week, my mother dragged me to the door of Miss Barrett's bungalow. Her knock betrayed how angry she was at the whole affair, though that had already been made perfectly clear to me. Inside, there was a tea table set for three with Marietta biscuits. We sat—me frozen with fear, my mother rigid and bristling. Miss Barrett poured the tea and related the Latin names episode, though she may have already done so in her letter. She told my mother she thought I was brilliant, and my mother stiffened to the point that I imagined she was about to shatter. She nodded her way through the rest of tea and then walked me home in a terrifying silence. When we arrived, she berated me for calling attention to myself. A smart woman would never marry into a family with a good estate, she told me, even though she herself was well educated. Men wanted someone, a chatelaine, to efficiently run their households and servants, not best them in arguments. I was to keep my mouth shut and avoid any repeat scenes that might embarrass her. Throughout her tirade, I half hid in my safe space behind the couch. When she finished, I simply nodded. We never talked about it again.

After two years apart, mostly it was my mother who decided to move back to England for a year, where we lived with my father in London long enough for me to be confirmed at Brompton Oratory in Knightsbridge. Then, when I was nearly twelve, my mother returned to Ireland with me in tow. I kept on doing my best to avoid attracting attention, both at school and at home, and found a way to live that worked for me and didn't irritate anyone. That camouflaged state became my default mode.

But then, within a year, whatever was still tethering me to the Earth loosened further still when I became an orphan.

The Yellow
Box of Paints

WHEN I WAS eight years old, my mother gave me a box of paints. It was the only present I remember receiving from her. She had found me drawing on the underside of the dining room table with coloured chalk; I was always drawing on stray pieces of paper, even bits of the newspaper. Terrified, I watched as she examined the art and then looked at me, saying nothing. The next day she went downtown and bought me a nice set of Winsor & Newton watercolours.

I loved that paintbox. I even abandoned my dolls for it. It was long and slim with a yellow top that folded backwards to form a flat surface that could be used as a mixing palette. The set came with one brush, but my mother had added a few more. One larger brush was made of camel hair. That was my favourite because the light brown hairs felt alive under my fingers.

I remember one day when I was about twelve I was out with my watercolours, painting some flowers. I needed fresh water, so I walked into the house carrying my jam jar full of discoloured water and soaking brushes. I was concentrating on carrying the jar carefully and, at first, did not notice my mother.

She was leaning with one arm on the mantelpiece, dangling a cigarette whose smoke was disappearing up the chimney. She was reading what looked like a letter, an official one, her eyes tracking every line. Suddenly she shouted "Jack," in what sounded like triumph. I stopped.

"Jack, the bastard, is dead," she roared, laughing as if she had won a game. I took her to mean what she said: *My father is dead.* I turned on my heel into the kitchen, where I washed my camel-hair brush with all the tenderness my young life could muster, pretending that the soft bristles were my father's hair. I never learned the circumstances of his death, and never saw him again.

———

I fell off my bike on the day of the car crash that killed my mother, just a few months later. I hit my head hard and suffered a concussion. When a neighbour found me, I was not sure where I was and the neighbour took me home. My mother was out visiting, and someone told me they'd called her to come home and that'd she'd be with me soon. I waited but she never came. As evening arrived I listened for the sounds of her high heels in the hall until I eventually fell asleep.

I was woken just before dawn, hints of light filtering through the curtains. A driver named Johnny Hayes had come to pick me up. I knew Johnny, who had driven me before, but he would not tell me where I was being taken in such a hurry or why. The streets were mostly empty and the car was moving faster than any I had ever been in, well past the speed limit. Sunrise tinted the clouds out the left-hand windows a soft rose. I leaned my head on one hand and stared at the sky as it deepened to crimson. *Your mother is dead*, the clouds seemed to say over and over.

When the car pulled in at Mallow General Hospital, Johnny did not even have time to pull off his driving gloves before I sprang from the car. I ran inside, veering past Emergency and into the left wing, following a gut feeling as to where I would find my mother. I kept on running, down a narrow corridor with wards opening to the left, at last ending up in a dimly lit room with a single bed. Going right up to the iron rails, I saw my mother. Each of her limbs and even her neck had been tied to the bed frame with torn strips of sheet. A clean white sheet covered her to her chin. The ties were stained as if they had been used before. The contrast of the grimy ties and clean white sheet spoke of horror to me, as did her skin, which was the colour of chalk. Looking down at her, I knew without a doubt that she had bled to death.

I bent down to kiss her cold cheek. To touch her, to feel part of her, to hold something that was vanishing away from me forever. As I lifted my head, the chief of surgery came charging into the room, followed by the matron and several nursing sisters, all of them barking at each other and me. "How has this child been let in to see her mother like this? It's monstrous."

I don't remember being led away. I don't remember being driven home again by Johnny Hayes. But I do remember that I desperately wanted my father, who was also gone from me for good.

———

In the wake of the crash, I was designated a ward of the Irish courts and made to appear before a judge in Cork whose job it was to decide, in his words, what to do with me. The typical fate of members of my wretched new orphan class, at least according to the Catholic Church, was confinement in a Magdalene Laundry, horrifying prisons originally built to "house" prostitutes and unwed mothers, which would decades later be exposed as nightmarish hotbeds of abuse and death. The decision to send me packing to a local laundry called Sunday's Well, according to the judge, should have been a simple one. My case, however, was not simple.

Wearing my green and slate-grey school uniform, I was led into the judge's chambers. From behind his huge, dark wooden desk, he spent the next several minutes voicing concerns over what would happen to him if he sent a Beresford to the laundries. Eventually, he got around to telling me that my Uncle Pat had offered to take me in and care for me until I turned twenty-one. Would I, he asked, be willing to live with Patrick O'Donoghue at Belgrave Place? I told the judge I would, which earned me a relieved smile from him.

The decision didn't erase the threat of the laundries completely, though. My freedom was dependent on adherence to a handful of conditions laid out by the judge in that first meeting: I would be required to appear in court every three months, so it could be determined that I was not going astray. My material needs—chiefly funds for school and clothes—would be administered by the courts out of my inheritance. I would also be given a curfew of ten o'clock. Violation of any of these terms, as well as, I suppose, my uncle getting sick of me, would land me in Sunday's Well.

As terrifying as that prospect was, I had more immediate concerns. Becoming an orphan hadn't made me any more interesting to Uncle Pat. He'd stepped up for me in the courts, and I was grateful for that, but that didn't mean he was prepared to do the same when it came to actually caring for me. He kept the same routine he had before becoming my legal guardian, and seemingly didn't spare me a thought. Perhaps because I'd been conditioned by my mother into near invisibility, I seemed to make it easy for adults to pretend I didn't exist. Though the house in Belgrave Place saw more people and activity in and around the funeral than it had in years, no one thought to ask me whether I was okay. Auntie Biddy, on whom I'd always been able to depend on for compassion and kindness, was in hospital at the time, being treated for pancreatic cancer; she would die a couple of years later. I basically curled up in a corner of the living room and stayed there. No one even thought to feed me.

I don't know how long I went without eating, but I know the funeral had come and gone so it could have been days. A friend of my mother's, Bridie Hayes, visited the house. Entering the kitchen, she asked the gathered adults where I was. No one answered, but she searched around and spotted me tucked into my corner and rushed to my side. "Has no one thought about this child?" she asked, turning and staring them all down. "Who's been feeding her?" Her questions were met with silence; I even thought I detected a bit of shame. With the rest of the adults frozen in place, Bridie set about making me scrambled eggs. It was the first food I'd had since the day of the crash and, oh my God, those scrambled eggs were the best thing that's ever passed my lips, better than any dish in the best restaurant I've eaten in since that day. I was so hungry and thirsty and, as I ate, Bridie gave them all hell in between bouts of scowling. She told them they were a disgrace for ignoring me. But once I'd finished, she left the house and everyone went right back to neglecting me.

Sometime later, I found myself alone in the kitchen and dreadfully hungry. I remember going into a bread cupboard and finding a skull loaf from Thompsons Bakery—a round, crusty bread with an *X* slashed into its top to prevent it from splitting as it rose. I was starving and my hand was small enough to fit into a hole I tore in the side. Bit by bit, I pulled out and ate the soft white centre of the loaf, hollowing it out so the crust stayed intact. Uncle Pat never said anything to me about that, though he must've found the empty husk of bread in the cupboard.

I'm not sure when or where Uncle Pat ate, too early for breakfast or too late for dinner, but during our first months together I can't remember a single meal we shared in the house. The lack of available food together with my grief took a physical toll on me—I mean, of course it did. On Sundays (it always seemed to happen on Sundays) I'd often conk out, fainting from malnutrition, and then be found on the floor. I was so depleted I began to contract strep throat every couple of weeks, it seemed. Though I was wasting away in plain sight, if anyone were to actually look, no one took measures to protect my wellbeing. I became as thin as a rail.

I went to school as usual, and had regular visits with my barrister and lawyer and interactions with clerks of the court and other functionaries. Clearly, none of them cared to notice my state. Uncle Pat's negligence, though, remains the hardest for me to understand, particularly given the version of him I would come to know and love in later years. My mother always said that her brother had his head in the clouds, possessing no understanding of the way families functioned. But to not realize that a child in his care needed to be fed? That remains a mystery to me.

Eventually, desperation led me to action and innovation. After I found a French cookbook, bound in linen, in the house, I decided to cook for myself. I asked Uncle Pat about the cookbook—he was always willing to discuss books—and he told me that it was my father's and its presence could be credited to my father owning a vineyard in Bordeaux. (Later I would realize he meant that as an explanation for why it was written in French.) I had seen enough cooking second-hand by that point to form a simple plan. I got out a pot and fetched four potatoes and washed them. The potatoes went into the pot and I covered them with water and set the lot on the gas stove. The cookbook, I thought, would provide the crucial missing piece: how long it took to boil a potato. Was I measuring in seconds, minutes or hours? I flipped the pages to find out.

The cookbook, it turned out, offered no discernible information about boiled potatoes. So I decided to discover the cooking time myself. Armed with a fork, I got the potatoes boiling, then every few minutes prodded them to see if they were softening. They started hard as rocks, bouncing away from me in the bubbling water. I prodded them again and again, still with the same result. I began to doubt myself—maybe there was some crucial step I'd missed. I stuck with it, though, closing the cookbook and, in an unconscious gesture of faith, returning it to its place on the shelf. Minutes later, I was able to puncture the skin with my fork, and after another few minutes, it went right through the potato. I had done it.

I turned off the gas, then drained the pot without any major burns and slipped the hot potatoes onto a cold plate. The skins were now opening and smiling up at me. I jabbed the fork in the air in victory. "At last," I muttered through a mouthful and finished off all four potatoes. Maybe, now that I was a self-trained cook, I could finally get a grip on how to survive my orphan state.

To the Valley

EVERY SUMMER THAT I lived in Ireland I had been sent to Lisheens. Uncle Pat saw no reason to interrupt that arrangement, so the day after school let out I was put on a bus at the central station and entrusted to the care of the loquacious conductor, Michael Murphy, for the two-hour ride to the valley. Our destination, Ballylickey, was a small village at the point where the Owvane River reached the coast at Bantry Bay. I climbed down the steps and out the bus's folding door, where I was met by another Patrick, this one my mother's cousin, whom everyone called Pat Lisheens.

Pat was around forty, a good-looking, sunny man with a quick wit, the "gift of the gab," and the wide, blunt hands and thick wrists of a farmer. He waited for me, as he did each year, sitting on the driver's seat of his horse-drawn cart, the reins lying slack over one knee. I hopped up with his help and then he drove me and my suitcase the few magical miles from Ballylickey up to the farm in Lisheens. The horse's tail swished amiably back and forth in front of us as we went, wafting its animal smell and the sweetness of fresh hay, which mingled with the fragrance of woodbine that hung in the air. I love horses, and being around one always made me happy. Perched way up on the cart and ferried in the fresh country air by such a beautiful creature, I felt for a moment like a fairy queen returning to her castle in the mountain mist.

There was a Catholic church three miles away in Kaelkill, which acted as a gathering place. The public nurse, Nurse Creedon, had a nearby office packed with mismatched wooden chairs that served as an examination room. But other than that, there wasn't really a "town" to speak of in Lisheens, just a collection of family farms and labourers' cottages on either side of the river. About a thousand people were living in the area and I was related to all of them in one way or another, though often only distantly. Pat lived with his mother, my great-aunt Nellie, on a forty-five-acre farm high up the side of the valley that had been part of our family's land holdings for centuries. His well-educated sister, Mary, was usually away working in England, which agreed with me because I didn't think she took to me much, at least at first. Nellie's husband, my great-uncle Willie, had died when I was a young child, so Pat worked the land alone, though I helped in the summers and he could count on neighbours to pitch in if a truly big job needed to get done. Nellie handled the animals—the pigs, sheep and chickens, as well as the cows. Both Nellie and Pat managed the dairy and Nellie did all the cooking and cleaning. Theirs was the valley's standard division of labour, and the two of them coaxed the farm into producing almost everything they needed to live.

There were many things I'd loved about the valley since my first visit. Lisheens has a unique place in the landscape of my mind even today. There was an almost tangible generosity of spirit in the air. I was fortunate to catch some of it in the time I spent with Nellie and Pat. The Brehon laws of hospitality were still as strong as they had always been, and according to those laws, as an orphan I became everyone's child.

Even the poorest of the poor felt that it was their privilege to give me something, if only a single ripe Bramley apple, or the choice gooseberry from their gooseberry bush outside their front door, or the first ripe strawberry.

After my parents' death, something deepened in my relationships in the valley. People looked at me differently, greeting me with a warmth that made tears come to my eyes. It was as if they understood that death was not a disease that was catching. Though I had no money, they treated me as if I'd inherited something of importance, as if I was suddenly a person with value. Nellie had always been kind to me, if in a slightly detached way. But from the moment I arrived that first summer after my parents died, she treated me as though I actually deserved to be cared for, as if I'd earned her attention just by being myself. I'd never felt that from anyone before and I fell head over heels in love with her. Every morning when I woke, I did my best to believe it wasn't a trick or a mistake soon to be corrected.

Of course, I was in terrible shape and desperately needed Nellie's care. She noticed that I looked like a stick. I crossed her threshold half-starved, not much more than bones and clothes, and I had trouble eating anything she put in front of me. Nellie soon sent out for a bag of Macroom oatmeal. A bowl of that stuff was like a restorative tonic, meant to stick to my ribs and give me a firm foundation to build my strength back up. She set a bowl of it on the table on my second morning and asked me to eat as much as I could. She also prescribed glasses of buttermilk, which I thought were awful. Each serving was flecked with small chunks of butter and wanted to come back up the second I'd gotten it down. Dutifully, though, I did as she instructed. I liked much better the bastible bread with currants she made for me as a treat. How long it took me to physically recover from those first few months of living at Belgrave Place with Uncle Pat, I'm not sure. But from the moment I arrived in Lisheens that summer, I stopped fainting on Sundays.

―――

The name Lisheens means much to people who speak Gaelic. It opens a doorway into a different world. First off, the name itself is ancient, though the British soldiers who came with their survey charts for colonization changed it. In old Gaelic, *Lios* means fairy mound or fairy ring or, from even earlier, the enclosed ground of an ancient dwelling place. The ending, "sheens," comes from *sí* in old Gaelic, meaning *aossí* or inhabitants of the fairy mounds. The valley is rife with stone artifacts from the time of the Druids, who were the elite educated class of the Celtic culture—the doctors and surgeons, astronomers and mathematicians, philosophers, poets, and historians. The hillsides were dotted with altars, ring circles, cairns, sacred stones, Ogham stones and holy wells. The turf bogs turn up treasures like baskets of butter, gold ornaments, or vessels of honey that have been preserved throughout the ages. When I was a child the valley might well have been the most concentrated, untouched site of Celtic culture in all of Ireland.

Though it was devastated by English occupation, the society that spawned that culture was once truly formidable. In the time of Christ, it had already spread from Ireland to fix roots in parts of England and Scotland. It travelled east through Germany and central Europe as far as Ukraine; and moved south from there through the Baltic states to Galatia of Turkey, and on a separate line from France to Italy to the coast of North Africa. It spread along the Silk Road as well. To this day, there is a part of central northern China that is Celtic in flavour.

The Celts had a written alphabet, Ogham, whose traces are still carved into those Ogham stones, that is thought to have originated as far back as the first century BC. Their society was governed by a set of democratically established laws, the Brehon, which changed over the decades to reflect the lives of the people. The Brehon Laws were codified in AD 438 by the High King, Laoghaire of Tara, who immediately established a royal commission to continually re-examine them and ensure they remained a true democratic representation of the rights of every man, woman and child in the Celtic world. The Celts who conducted this re-examining were the Brehon judges; they carried out a form of jurisprudence practised in Ireland for more than thirteen hundred years. My maternal grandfather, Daniel O'Donoghue, was one of the last Brehon judges to have his *breithiúnas*, or judgments, put in place.

The O'Donoghues were aristocracy and the family seat, the Castle of Ross in Killarney, was one of the powerhouses of scholarship in the Celtic world. My grandfather had grown up in a family illustrious enough to have Irish servants, which was very unusual, but he'd fallen in love with one of the maids, my grandmother, and ran away with her on horseback. He refused to speak English, signing their marriage licence with an angry *X*, but he spoke and wrote Gaelic, Latin and Greek—the languages of the Celtic elite. His Gaelic was so good the University of Cork dispatched scholars to transcribe it. He was known to have *blas*, which rendered in English means something like his speech had "a sweetness on the tongue." Along with a special facility with language, he had an encyclopedic knowledge of poetry and literature, law and history. He was a living library, a human repository for all of the cultural knowledge the English tried to eradicate through occupation and the Penal Laws.

My mother and her siblings were raised in a place called Lackavane, which was a smaller valley, hidden high among the Caha peaks directly above Lisheens. Lackavane was a natural fortress, relatively safe from attack by the English soldiers who enforced the Penal Laws for centuries, and it was also protected by rock faces of slate that would sheer off and cut the fetlocks of the English horses. Guerilla raiding parties from Lackavane could hit a group of soldiers and then disappear into the mountain mists, beyond reach. It was also a place where ancient knowledge was stored for safekeeping. As a result, the Celtic culture was better preserved and protected in Lackavane and Lisheens than it was in much of the rest of Ireland. And owing to our bloodline and my grandfather's status, my mother's family was the most prominent in the area.

Where England had failed, though, Western consumer culture with its urbanization was making easy strides. The older inhabitants of Lisheens still lived much the way people had in the valley for hundreds of years. They were subsistence farmers, who occasionally made use of a tractor but mostly stuck to the old ways of working the land. They were content with their lot in life, their poetry, songs, music and language and their place in the world. But their children and grandchildren had sampled the modern age and were eager to enter it as fully as they could. They wanted money, a car and a new life in America or England. Those who stayed had heard of the newest machinery and chemical fertilizers and wanted to double or triple their yields to make more money. The feeling of wanting was palpable in the valley, and it had the power to lure people away from the traditions they'd grown up with. Everywhere in Ireland, the younger generations were turning away from the ancient knowledge, casting it as little more than superstition. But in Lisheens, the older people were still rich with what the young generation didn't want.

I was the last of the O'Donoghue bloodline entitled to the ancient inheritance my grandfather had kept safe, and this was the world I had now entered, the world of Lisheens.

——

Aunt Nellie, my mother had told me, was once the most beautiful woman in the south of Ireland. She was in her early sixties when I was orphaned, but it wasn't hard to believe my mother's assertion. Nellie had a lovely high forehead, clear blue eyes and a noble nose. Her skin was unlined, her cheeks still flushed with rosy health, and her hair a shimmering silver that fell to her waist but that she always wore whipped up and held in place by a tortoiseshell comb. She attached a silver safety pin to her blouse, above her right breast, and usually paired the blouse with an old-style handwoven skirt—linen in the summer and wool as the air cooled—that reached down to her ankles. She had a gentleness to the way she did everything, even when her task was a difficult physical one, like churning butter, and she always asked me for help with such softness. "Diana, could you help strain off the potatoes?"

It was from that gentle voice that I first heard about the plan for my Celtic education. Usually the transfer of knowledge in the valley from one generation to another happened only within families. While you worked alongside your father or shared a meal with your aunt, they passed on the wisdom that had been given to them and added anything they'd learned in their years of life. But I was a different story. First, because my grandfather had been the keeper of a vast store of knowledge and, as his blood, I was entitled to the same. And second, because under the Brehon Laws, an orphan became everyone's child. This next bit is guesswork on my part, but I suspect that when Nellie saw the state of me when I arrived at her door, she called together the oldest generation of women in Lisheens, most in their eighties and nineties. She put it to them that I was their responsibility by ancient law and I was wasting away to nothing, and they agreed to do something about it. What she told me was that I was going to complete what she called a Brehon wardship, an arrangement in which I would be taught the things I needed to know to be able to take care of myself as a young female growing into womanhood. I would have many teachers. Each would send for me when it was their time. She would be the first of them and our lessons would begin immediately.

The people of Lisheens ate only two meals in a day, breakfast upon rising and dinner around two in the afternoon. There was no lunch, which was an English custom. After dinner, when the dew was off everything, Nellie told me she was ahead of her work for once. "I find myself with some time on my hands," she said, and proposed that we take a walk. This was the first lesson of my wardship.

Leaving the cool of the kitchen, I followed her past the flowering shrubs, the *Hebe* with their long tails of blue stars given to her as a wedding gift. We walked down the path side by side to the gate where Nellie paused, as she always did, to admire the Caha range—her mountains. Closing the gate behind us, we turned not down the slope towards the valley's main road, but up along the track that led to my weeping stone. We passed the stone and kept going around the top of the field, eventually coming to the edge of an old earthen ringfort where we had views of the entire valley.

What remained of the outer wall sat on a bank that had been cut into the hill like the edge of a terraced rice paddy. Ringforts were protected—never touched by the plough. The wall and its bank curved away in either direction, eventually looping around to reconnect somewhere out of our sight. In front of us, the ditch created by the bank was thickly blanketed with plants, and it was there Nellie led me. As we neared the ditch, I began to be able to pick out specific features from the riot of green. I saw leaves of all sizes and shapes; runner plants carelessly growing over and around their neighbours; the fragile face of a flower poking through here and there, trying to stand its ground. Nellie, calm and confident, reached into this mess and picked a single leaf. Using her thumb like a pestle, she ground the leaf into her palm and then held its crumbled green body to my nose, which was immediately filled with a fragrance like spearmint. "This is pennyroyal," she told me. "You won't forget its smell." She picked another leaf from the same plant and offered it to me. "Remember how it looks," she said, and I studied the small, elliptical leaf in my palm. Then I tried to recreate it in my mind—its exact shade of deep green, the mauve-blue flowers and the furrowed veins extending from its midrib.

After a moment, Nellie continued my instruction. "This is an important medicine for so many things," she said. "It can be smoked or used fresh. We take it to cure colds in the winter and use it to keep insects away and treat their bites in the summer. It is an old, old herb used in ceremonies long ago. Its medicine is strong and if you forget the look, you can always recognize it by its smell."

Letting the bruised leaf fall from her palm, she scanned the ditch and reached in to pluck something else. Then she set in on an explanation of its medicinal properties. This was how my first medicine walk progressed; the ditch was Nellie's pharmacy, and she highlighted species after species. She had a cure for diseases I had never heard of—mental illnesses, digestive upsets, heart problems and skin troubles. The further she progressed along the ditch, the more overwhelmed I became. *How, in the name of God, was I possibly going to remember everything?* Having sat in front of a judge while my fate was being decided and then having spent months on the verge of starvation, I knew I needed advice on how to survive. I listened as carefully as I could, but there was just so much to know. Right as I started to give in to despair, Nellie noticed the look on my face and dropped the stem in her hand.

"Well, Gidl," she said, using the endearment she and Pat had adopted for me, "that's enough for today."

I was sure she was stopping because I'd failed her, and I couldn't keep my dismay from writing itself across my face. Noticing, she laid a comforting hand on my shoulder. "I know there is so much for you to learn, but I'm thinking you're up to it," she assured me. "Repetition makes a miracle of the mind. Don't worry, you'll do it bit by bit by bit by bit by bit."

And that is exactly how my learning progressed: bit by bit by bit. Lessons from Nellie or occasionally Pat were interspersed with lessons from teachers flung all over the valley. My "faculty" ultimately numbered over twenty people, and Nurse Creedon, who knew everyone in the valley, was the organizational force that scheduled them all.

Nurse Creedon had grown up in Lisheens. In her twenties, she'd left Ireland to study in New York before returning, degree in hand, to serve her community. Unlike others who'd had a taste of the modern world, however, she didn't reject the old ways. She consulted all the people of the valley for their knowledge of medicinal plants and she wasn't shy about using natural medicines. Though her official title was "public nurse," she effectively functioned as an MD whose pharmacy was the valley and community itself. She kept her surgery near the church, where many people came for mass every day. There were no phones in Lisheens, so after mass, anyone who had a bit of knowledge they wanted to share with me would pay Nurse Creedon a visit, enjoy a cup of strong tea or Camp coffee, and let her know. Each offering was equal to any other: a small bit was as good as a big one, since they all contributed to the pot. She would then send a runner—a long-legged, big-lunged boy—up to Pat at the farmhouse, who would then pass along the invitation to me. This whole wonderful chain was largely invisible to me. The bit of it I encountered was usually a simple line from Pat in the morning, something to the tune of, "Diana, such-and-such a person up in the mountains or over on *Cnoc Búi* wants to see you."

The lessons always took place in the afternoon. I would make the journey to wherever I'd been invited and walk through the door of a farmhouse. My host, sitting at the kitchen table, would sing a song of ancient times in Gaelic to passage me into their house. They'd offer a cup of tea, which would be prepared and poured as carefully as in a Chinese tea ceremony, and then I'd sit and they might flit around the room reciting a line of poetry, or a few—the beginning of a process of bouncing the Gaelic language back and forth between us.

I was taught the medicines of species, as I had been on the ringfort walk with Aunt Nellie, and any number of practical skills. For example, one of the women up in Kealkill was a butter expert; she produced varieties of butter the way we think of artisans creating cheese or wine. I was brought to her to be shown different kinds of butter that came from different kinds of *terroir* and their uses—how, for instance, a coating of butter could prevent eggs from oxidizing, keeping them fresh for a long time without refrigeration.

I remember the taste of her butters. One orange-coloured butter was salted and strong, and at first I didn't like the flavour of it. Then I got used to it. She told me that the colour came from the first summer feeds when the cows were released into the high summer pastures. That was when I realized that variety or biodiversity was a good thing, even for a cow.

Many of the lessons, though, especially early on, were psychological. My teachers knew I was in danger, that I would face threats as an orphan and as a woman. They also knew that I was dealing with immense grief and trauma. So they taught me how to process the pain of life and care for myself in body, mind and soul. They believed that I could accomplish anything, but that to reach my biggest goals, I would have to start with believing that I could. So I was taught to love myself and trust in my own abilities.

Learning to value myself was particularly difficult. I was coming off fourteen years of self-nullification and many months of horror that had left me feeling cursed. But even in these more abstract lessons, my teachers provided clear directions and practical tools. With the tea on the table, one of them might say, "Sit down now, Diana, and give your mouth a rest. Well, you know, we were talking and we were worried that maybe you might get lost." Not physically, but mentally—that my sense of self or purpose would abandon me. "So, I'm going to tell you how I manage when I feel lost or overwhelmed."

Through their own experience and practice, they taught me to "go into silence," a Celtic form of meditation. They might ask me to think of a chair. I would call one up in my mind, usually one of the wooden chairs in Aunt Nellie's kitchen. When I had it fixed, they'd tell me to enter the place where that chair sat. I had to work at it, but freeing myself from worry and irritation so I could calmly sit in that mental chair left my whole body feeling rested and gave a holiday to my brain.

Another time, I was asked to remember the happiest day of my life, to reconstruct it in my mind. Then I was taught to enter that same feeling of happiness, like walking barefoot in the clean sand of an ebbing tide. Afterwards, I would be able to go back and visit it whenever I needed strength, to be cheered up or just to feel really good for a while.

In my happy place, I was lying on a mattress stuffed with fresh straw in the front bedroom at the farmhouse in Lisheens. Morning sunlight was falling through the window and bathing my feet; I had hot ankles and it was a lovely feeling. The straw was heavenly and sweet and its smell was rising all around the bed. I could hear the chickens clucking outside, and there were cows off in the distance and horses standing closer to the house. I knew I could go down into the stables and get a fresh brown egg for my breakfast. I was anticipating the taste of that egg, and it was like a God-given time—a special, special time. I still go back there in my mind and find myself wearing that joy again.

————

Nellie and Pat's home was a traditional Irish farmhouse, two storeys tall and built from fieldstone with a grey slate roof in the old style that would've gone back at least a couple of hundred years. The farmhouse, dairy and stable were whitewashed. The dairy, where Nellie prepared the milk, cream and butter, was built onto the back of the house. The stables, attached to the east side, held the horse-drawn equipment and all the leathers—saddles, harnesses, blinders, bridles and bits that spilled like a waterfall down the wall they were hung on.

Tucked safely under that cascade of leather were the handwoven laying baskets, packed with soft beds of hay and straw. The hens would leave their chicken house and pick and flap their way to the stables to lay their eggs. Each of them had a particular basket she preferred; if I picked one off her usual spot and moved her to a new basket, she'd always return to the one she'd chosen herself.

I had a favourite hen, my queen. Her eggs were big and brown, sprayed with constellations of darker brown freckles. Every morning, I would go out to the stables and retrieve the offering from her basket. That, along with a slice of bastible bread, was breakfast.

One morning, I was sitting on a rough-hewn wooden stanchion in the stable examining my newest egg, when I heard the unmistakable sound of fully shod hooves clipping along the flagstone path that ran around the milking shed. The prospect of meeting some new horses had me out of the stables and across the yard in no time. My feet barely touched the ground until I laid eyes on them. They were a pair of stunning chestnut mares held in place by leather leads secured under a large, round stone. I stopped dead to take in the powerful muscles of their flanks and legs and their gleaming coats. As they flicked flies with their tails, it dawned on me: they were racehorses!

I was so excited I said the word aloud—*racehorses*—and then I saw the smoke rings drifting up lazily from the other side of the mares. I crept slowly around to get a peek at their source and found that they originated from the short, curved pipe of a man who was sitting on a milking stool examining the horses' legs. He was about Nellie's age, well dressed in a three-piece suit of fine Irish tweed, with a gold watch chain hanging in a smooth arc against his vest and wild hair the colour of summer butter standing to attention on the top of his head. I had never seen hair like that and, as I stared in wonder, he turned to me.

His bright blue eyes froze me in place as he took the stem from his lips and let slip another perfect procession of smoke rings. Then, he said, "You the Gidl?"

I couldn't speak, so he asked again. When still I stayed silent, he softened his approach. "I'm sorry, *leanbh*," he said, using the more affectionate word for a child. "Don't take any notice of me now. I have a job of work to do here with these horses. They have been ridden too hard at the Killarney races, and I have to fix them up. First, though, I have to find what's wrong with the pair of them. Come here and help me if you'd like. We'll talk to them together. Just watch out for the hind legs—they're fast with a kick."

He rose from the stool, dusting himself off with one hand and slipping the pipe into his vest pocket with the other. It finally dawned on me who he was: Nellie's brother, my mother's beloved uncle, Denny O'Donoghue from Lackavane. Uncle Denny was a *cúipinéir*, a bonesetter—one of the last bonesetters in Munster. Famous for his gift with animals, he was something like a vet, only better. My mother had told me he could communicate with horses, whistling and whispering just what they needed to hear in soft, soothing Gaelic. He could glean information from an animal's coat, and knew how they would best be fed and run and spoken to. People came from far and wide for the magic he seemed to have at his fingertips, and he was on call at tracks and barns across the south of Ireland. Though he was very handsome, he'd never married. I don't think he got fussed by women too much. He concerned himself with horses.

"Uncle Denny," I said, breaking my silence at last with his name, spoken somewhere between a statement and a question.

"Well, yes, that would be me," he answered with an easy smile. "They tell me the women from here up to Kaelkill are taking you on. Well, Gidl, I have a bit to add myself, so I do."

Denny took one of the horses by the halter, liberating its lead from under the stone with his shoe. Gently he led the mare apart a handful of paces as she whinnied in pain. She had a limp in her left hind leg and didn't like to put her weight on it. "Easy now," Denny said through his teeth. He spoke directly to the mare, though I could tell much of what he said was for my benefit. "I know you're hurt but I think I can fix you. It's the tendon; it has heat in it. I think a little massage and finger-fixing will make you as good as new. Then I'll leave you and your sister here out to grass, so I will. The calcium in the ground will fill in any hairline cracks in your bones, and then we'll see how you feel."

Even in motion, Denny had a stillness about him, the calm sense that he never hurried anything. His ease was amazing, and inspiring. He could sit and stare into space and think, just be alone with himself in total comfort. Leading the mare a few more feet, he continued to speak to both of us. "Easy now," he said. "You'll have to stand for me a while, so I can take stock of you and see that there's nothing else wrong."

He looked her over again and then, calming her with his voice, massaged her left flank, moving his hand again and again from her hip down to a point just above her hock. "We're getting there," he said after a minute or two. "I think the knot is loosening, unless I'm much mistaken. The tibia and fibula are stretched a bit, but they're bones and they will go back into place with a nice little rest."

He asked the mare, "Do you hear me now? A little rest—no galloping for the next week. I'll be here and I'll have both eyes on you day and night. That hock will mend and you'll be your same troublesome self again in no time."

As he released the mare and her sister into one of the side pastures, I could see that already she was moving better. Her limp was far less pronounced, and she'd stopped letting out the sad sounds of an animal in pain. My mother had loved Denny deeply for his connection to animals, especially horses, and already I felt a similar affection blooming. As he brought over a second milking stool and got us settled, side by side, to keep an eye on the mares, I worked up the courage to show him an affliction of my own.

For years, Pat and Nellie had been chastising me for spoiling the animals. I handed out goodies and affection freely, and as a result, I was often trailed by a flock of chickens or ducks, the pair of dogs, Flossie and Flo, some cats, some sheep or any combination thereof. In addition to my favourite laying hen, I adored a huge Landrace sow called Daisy who would come running when I called her. Daisy would lie down and I'd scratch her belly, and a handful of rough words would be lofted my way by Pat because I was making her into a pet. I also tended to get shouted at when I fed the calves. I stood on the gate, under an arch covered by a beautiful, climbing Lady Banks rose with rare pink flowers and a stem like a tree trunk, and passed hay through the fence. Then I'd let the calves suck on my hands, which felt like they were about to take the skin clean off from my wrists to my fingertips, but was somehow also nice.

Pat and Nellie's admonishments had always been delivered with a note of humour, enough warmth in their voices to let me know that what I was up to wasn't all that bad. But in the week before I met Denny, a line of ugly warts had developed across my knuckles. I was deeply ashamed of these warts. It was scary the way they grew bigger by the day. But more importantly, I was sure I'd gotten them from mishandling animals. I had done something I wasn't supposed to and there was the proof, plain for all, especially me, to see.

I had done my best to hide the warts from people, while also trying to make sure they didn't spread. But I decided to show them to Denny. He took my hands in his and I saw no judgment as he appraised them. He didn't scold me—he just told me he knew of a simple cure. Denny's treatment involved halving a Kerr's Pink potato fresh from the field and scooping out the flesh in the middle. He put salt in the makeshift bowl he'd formed and told me that overnight the salt would draw water from the potato, creating a little pool. "Take that water and rub it on the warts," he instructed. "Use a new potato every day and repeat what I've just shown you. After three weeks, you'll be fully healed." I followed Denny's instructions and treated my knuckles every day. By the time I was to return to Cork, the warts were gone.

This may seem like a small thing, but it was powerful to see the effectiveness of Denny's cure. I was frightened of those warts. They were the latest misfortune in my doomed life and a part of me believed I'd never be rid of them. And then they were just gone. It was magic and, at the same time, concrete proof of the truth behind the lessons I was being taught in Lisheens and of the awesome powers of plants. I was fascinated. That was the start of my being totally swept up by excitement over what else these elders had to teach me. I couldn't wait for my next lesson and the bit of ancient magic it held. But I would have to wait, because Cork, school, winter and Belgrave Place all beckoned.

No Burden for a Woman to Be Educated

RETURNING TO CORK after the first summer of my wardship felt like another loss. The city was such a lonely place for me. I was alone in the middle of a crowd. I couldn't help but grieve the absence of Nellie, Pat, Denny and my other teachers; of Daisy, my queen hen, Flossie, Flo and the other animals; of *Cnoc Buí* and the Caha mountains; of the sea and the sky; of the heather, gorse, holly and blackberries; and of course, the loss of Nellie's cooking. But even after just those few months of lessons, I was better equipped to deal with my loneliness. When the question popped into my head—*Why did all of this have to happen to me?*—I now felt I had something to respond with, not an answer but rather acceptance of the fact that an answer was unnecessary.

The valley had taught me that there is a reason for everything. I could trust in that and be comforted by it without knowing the reason itself. I was destined to face my portion of suffering, and while mine might seem larger than others', I now felt that I was also destined to outlast that pain. The warm wishes of Lisheens, the kindness I'd experienced there, walked with me and cloaked me. When I went to school, it was still there wrapped around me. It helped me feel whole and more able to simply be myself.

My schooling had been a point of contention between my parents. My father wanted to send me to an elite boarding school in the south of England, and he'd gone so far as to register me there mere days after I was born. My mother, however, insisted on an Irish education—in particular that I learn the Irish language. She ultimately won out. Before she died, she had enrolled me in a private school for girls in Cork called St. Aloysius School, run by the Sisters of Charity. Her will made special provision for my tuition there, and I was slated to begin that fall.

Little else about my situation was that clear-cut, particularly when it came to my finances. A week before school started, I was scheduled for another appearance in court to allow me to request money for "necessities." Food wasn't included in that category because the authorities assumed Uncle Pat was providing for me, but much of the money given to me for clothing and "sundries," the court's prudish word for my underthings, ultimately went instead to my nourishment. At fourteen, I began to do all the household shopping, as well as the housework and cooking. I would go to O'Flynn's shop, the local butcher in the English market, and ask for, say, a pound of lamb, and Mr. O'Flynn would wrap two pounds in paper, only charging me for the one. He never came out and acknowledged that I was in desperate straits—I was, after all, Rocky Donoghue's niece—but that discreet kindness helped to keep me alive. With these doubled provisions, I prepared regular family meals for myself, Uncle Pat and, while she was still alive, Auntie Biddy, which helped to bring us closer together. Though Pat never once asked why or how I'd taken it upon myself to cook, he eventually did start leaving grocery money on the kitchen table.

My relationship with Pat improved dramatically that fall. Lisheens had given me a spark of self-confidence, and I no longer felt such a pressing need to be silent, to hide central aspects of myself, like my intelligence. Pat, it turned out to my surprise, didn't share my mother's views on women's education. Far from it, in fact. He would, over the next few years, say to me again and again, "Diana, it is no burden for a woman to be educated." From our first days back together, he took obvious pleasure when I shared with him something I'd learned or introduced a topic for us to mull over or debate. One of the first instances I can remember was around Uncle Denny's cure for my warts. Pat had heard of the cure and told me it was an old one, but then at his bidding we began to investigate why it worked. *What types of potato were used? Did the chemistry vary between varieties? Did the application have to dry on the skin? Why did the potato need the sunlight of the window ledge?* Both of us contributed questions, and Pat took as great a delight as I did in discovering the answers. We found that any potato could be used, but it had to be a new potato of the same variety every day, and the action of the sunlight was crucial, because it produced solanine, a chemical that was an anti-viral that would kill warts on the skin.

He also encouraged me to read freely from his library. On the shelves at No. 5 Belgrave Place, Pat had amassed a collection of some ten thousand first editions. These books ranged widely in subject matter and they were joined by others, piled on the floor and waiting to be inspected. One of the first volumes I remember pulling down contained pages and pages of watercolours by the American painter Andrew Wyeth. Uncle Pat saw me sitting with it and asked what I was after. "You've got so many interesting books here," was all the answer I gave him, and that became our shared language: we both spoke books. We loved the design, the weight of the book in the hand, the fonts. A green row of Penguin books was one of his delights.

Two nights of the week, I took evening lessons at the Cork School of Art, sketching or painting right up to the edge of my curfew. But the other five, after I got home from school or from the tennis and camogie (the female version of hurling) I played on weekends and prepared dinner, Uncle Pat and I sat ourselves down in a pair of armchairs and read until bedtime. He always favoured the left-hand side of the fire, while I liked the right, nearer the long Georgian window. Occasionally, I stretched out on the red carpet like a contented cat. If he stumbled across a poem that struck a chord or encountered a bit of interesting information, he'd share it out loud. When he started to get hoarse from his recitation, I'd find a passage that jumped out in whatever I was reading and it would be my turn to read aloud. We passed most evenings like that, bartering the best passages back and forth. I found that I had a powerful thirst for knowledge, and sometimes felt on those nights that I was taking it in by osmosis—as though just being around that many books allowed me to absorb information through my skin.

At St. Aloysius, as at my previous schools, the other girls largely avoided me. The nuns seemed to know I was bright and to appreciate me for it, but they didn't show me any affection. They did, though, guard me in a quiet way. I suspect that because of their interest and possibly even intervention I was never teased or bullied. But without anyone to guide me or explain how things worked, I got into a scrape or two. Towards the end of my last year at St. Aloysius, I'd picked up somewhere that our finals, called the Leaving Certificate examination, were honour exams, honest tests of what you'd learned that year. Always striving to do the honourable thing, I didn't study at all for the tests. The morning of the exams, I got to school and all the girls

were sitting on the stairs with their books open, cramming in whatever would stick in their minds right up until the last minute. I was
scandalized and thought they were cheating until it was explained to
me that studying was the proper way to prepare for exams. I entered
the classroom terrified that I was about to expose myself as an idiot,
but I aced the exams despite myself, coming first. Next time, I knew
to study.

My loneliness wasn't always a bad thing. Knowing that I would be
ignored and friendless almost regardless of how I behaved freed me
to be more fully myself without fear of consequences. Scholarship was
a hidey hole, a place where I could escape negative feelings. But as I
learned more about the world, I realized that a life buried in books
quite suited me. When I had my first brush with hard science, that pas-
sion intensified. Here was an excellent reason to be at school, friends
or no: the sisters held the key to knowledge I was longing to unlock.

———

In my pleated green skirt, slate-grey blouse and tri-coloured tie, I sat
with the rest of my classmates in the school's chemistry lab waiting for
Sister Mercedes to appear. My blazer, with its bright crest, hung over
the back of my stool and there was a smell of gas in the air, evidence
that someone couldn't help but fiddle with the stopcock on their Bunsen burner. On the tables in front of every girl in the room—placed
there with care, I imagined—was a new chemistry textbook from an
American press, its spine uncracked, a very handsome volume that I
knew Uncle Pat would covet. I was busy running a finger admiringly

down the edge of the cover when the sister entered the lab with a portfolio of documents in her arms.

Our teacher set her things down on her desk and moved to the head of the classroom. She smelled the gas, her face scrunching in displeasure, but chose not to mention it before getting down to business, instructing us to open our books and read the first chapter in silence. There were groans and sniggers as Sister Mercedes took her seat, adjusted her wimple and proceeded to shuffle papers about with obvious purpose. I delicately opened the cover, pausing briefly to enjoy the soft crackle of it becoming my textbook, and then let the rest of the classroom melt away as I trained all my attention on the words in front of me.

I began at the copyright page, noting the name of the publishing house before moving on to the assigned chapter, which was on the properties of hydrogen. The pages were crisp; full-colour illustrations sang out from them. Hydrogen, I learned, had a singular electron valance of one and an atomic number of one, an atomic weight of 1.0079. Its chemical code was the letter H. Everything about this first element made such sense to me. The sensation was similar to the one you have when you're reading a novel and see a feeling perfectly described that you only realize you have felt yourself after encountering the description. *Ah yes, of course. That's it!*

Because I was so excited, I read as fast as I could. The chapter was about ten pages long. When I came to the end, I put up my hand and coughed for the teacher's attention.

"Go on," she ordered. "Do the next chapter."

Oxygen. Atomic number eight, atomic weight of 15.9994—a lot heavier than hydrogen. Symbolized by the letter O. I was fascinated. The textbook to me was more engrossing than any comic. The gas leaking from its Bunsen burner could fill the lab, spark and blast Sister Mercedes, wimple and rosary beads singed, clear across the road and through the roof of her convent, and my eyes would've remained glued to the page. I reached the end of the chapter on oxygen and again raised my hand, though I coughed more softly this time.

The nun looked up from her papers. When she caught sight of me, a flash of indignation lit her face. "No!" she shouted. "You, young lady, are playing games!"

I looked over one shoulder and then the other to make sure that she was addressing me. All of us were so shocked that there was no sniggering or whispering. The room was dead silent. Sister Mercedes stood up slowly and dangerously. She closed her portfolio with great deliberation. Then she closed her eyes and fingered her beads, pausing on the large crucifix. She pressed the body of Christ with one finger as if it were a button capable of triggering the correct response. She opened her eyes. "Go on," she ordered, her voice dripping sarcasm. "Recite what you've read. We have all day."

As I felt the eyes of the room shift from her back onto me, all the joy I felt evaporated. I'd merely done what she asked, yet I could tell that I had somehow dug myself a deep hole. "I'm sorry, Sister Mercedes," I began, but she silenced me with a raised hand. "Stand up, Miss Beresford," she said, her voice still acidic, "and show us what you've learned."

With no idea what else to do, I began at the copyright page. I quoted from memory the name of the press. I added the date of publication for good measure and then I started in on the hydrogen. I tried to infuse my words with the fervour of prayer, hoping that would sway her. I recited page after page, including the page numbers. The rest of the room stayed silent. The nun's face was an unreadable mask. Later I would learn that

I have a photographic memory for equations, and a near-photographic memory for most everything else. I would also learn that not everyone's mind works the way mine does. In that moment, though, I knew only that I'd been asked to recite what I'd read. I started to relax a little as I moved into the chapter on oxygen, enough at least that I could again take pleasure in the disparity between the two atomic weights. I was partway down the left-hand column on page fifteen when the sister again held up a hand to silence me. I am almost sure that she swore under her breath before gathering her beads in one hand and walking out of the room. Moments later I saw her out the window, crossing the gravel courtyard and entering the convent.

What had I done? I was terrified. None of my classmates spoke or moved. There was a charge in the air, though I couldn't tell whether it was them sharing in my fear or anticipating an entertaining scene. For my part, I was sure I'd gotten the whole lot of us in trouble. I would be expelled and thrown in Sunday's Well, regardless of the judge's concerns about reprisals from the Beresfords. And all because of hydrogen and oxygen.

I was still lost in this thought when a girl scouting at the window announced in a loud whisper that Sister Mercedes was returning across the courtyard and "Bona" was with her. Sister Bonaventure was the headmistress of St. Aloysius. I was doomed.

The nuns whooshed into the lab and proceeded to the front of the room, their beads banging against the tables in their haste. I was told to stand up; my knees almost buckled under me, but I made it. "Do it again," Sister Mercedes told me. "The chapters. From memory."

Once more I started with the title. My hand lay on the closed book in front of me, the contact a reassuring anchor, maybe the only thing that kept me from fainting or floating away. Around page thirteen Bona told me to stop. I did, but remained standing just in case I was asked to repeat.

Bona turned to Sister Mercedes and the two shared a look I couldn't decipher. Then she turned back to me. "Sit down," she said.

I did and she left the chemistry lab without another word. Sister Mercedes stepped back behind her desk and settled to her papers after she directed us all to resume reading. Nobody said anything to me about the incident, neither my classmates nor my teachers.

A week later, in the middle of Sister Mercedes's class, I was asked to go down to Bona's office. Waiting for me there alongside the head-mistress was a man she introduced as Mr. Holland, a professor from University College Cork. "Miss Beresford," Bona said, "Mr. Holland here will be instructing you in mathematics for the foreseeable future."

I can chuckle now thinking back on my reaction, but in that moment I was ashamed. My lessons with Mr. Holland would, over the next three years, allow me to dive into university-level math ahead of my peers. Instructing me one-on-one in a separate classroom, Mr. Holland pushed me at a breakneck speed that couldn't have suited me any better. Under his tutelage, I devoured mathematics, sharing what I'd learned with Uncle Pat in the evenings and feeling more connected to my true self than I had at any other time within the walls of St. Aloysius.

But standing there in front of the stern headmistress and this man I'd never laid eyes on before, it was impossible not to flicker back to the mode of thinking my mother had ingrained in me—to be certain that being singled out in this way was proof that I'd done something terribly wrong.

The Meaning
of My Wardship

ALL DURING THE first summer of my wardship in the Valley of Lisheens I had remained a bit suspicious of the attention I was receiving. First, I doubted the motives of my teachers and assumed they'd quickly grow tired of me. Then I began to suspect that there must be something wrong with me for people to take such an interest and even a liking. I kept waiting for this defect to be revealed, to sabotage the kind course that had been set for me and aim me back towards misery. Only when I left Lisheens at the end of the summer, still wrapped in that cloak of warm wishes, did I finally start to seriously wonder whether my suspicions had any foundation.

Over the winter in Cork, I gave the situation a huge amount of thought. As I prepared dinner or cycled to school, or lay in bed waiting for sleep, I pored over every interaction I could remember from the valley, looking for cracks and fault lines. Instead, I turned up small, seemingly insignificant moments that, when examined in the cold city light, held evidence of empathy and care for me. I found further reasons to trust my teachers and their love. In the early summer, when the bus and horse-drawn cart brought me back to the valley and my lessons resumed as though we'd only left them the night before, the consistency provided the last proof I needed.

As before, my teachers didn't assign relative values to the knowledge they taught me. Nellie's huge inventory of plants and their uses was no more important in their eyes than my being taught to preserve a single egg in butter—I needed to learn it all. That esteem, which was heaped in equal measure on even the smallest tidbits, stories, songs and poems, is a crucial component of Celtic culture. And there was a constant reminder of it in every farmhouse from Ballylickey to Lackavane and beyond: the bed kept for the *seanchaí*.

The *seanchaí* was a wandering storyteller, a man with a prodigious memory and compelling delivery. His was an inherited position, passed down a family line of storytellers hundreds of years long. In the colder months, roughly from harvest to planting, he would travel from place to place sharing his stories with the people.

Nellie's kitchen was the biggest room in the farmhouse. It had a floor of beaten earth, hard and smooth as polished concrete, and a large wooden table with many chairs. Built into the western kitchen wall was a big, open, arched fireplace with a hob for her pans and bastibles and a hook for her three-legged pots. And against the wall by the front door, facing the fire, was the *leaba shuíocháin*, the settle bed.

The *leaba* was capable of seating two or three people or sleeping one with a cushion under them for comfort. The *seanchaí* had the right to that bed in every house he entered. He would send a runner ahead to warn of his arrival, and the following morning you'd come down to find him curled on the *leaba*, his coat bundled under his head for a pillow and his shoes drying in front of the fire. That night the kitchen would fill with neighbours and the *seanchaí* would unleash his arsenal: short, sharp stories called *gearrscéal*; beautiful poetry; ancient adventure stories filled with fairies, banshees, shapeshifters and the heroes of Irish legend; genealogical histories that revelled in the magical rope of humanity that bound us all together.

To the Celts, story is everything, both an act and a piece of creation. The *seanchaí* had the ability to spin whole worlds out of nothing, and people would travel miles to crowd in around him and the kitchen fire. I remember one evening just before I left for Cork and school. The *seanchaí* had chosen to stay at Nellie and Pat's (which would happen maybe twice a year) and an elderly couple called Patrick and Mary O'Flynn arrived at the door around dusk. Both were barefoot. They had taken off their shoes and homemade woollen socks to cross the streams on the way over from their farm. Their shoes were laced together and slung over their shoulders and their pockets bulged with their socks. Each had a walking stick and wore a colossal smile. Making it over to hear the storyteller was their triumph. With his hair curling wildly from the top of his head and dressed in his weathered and torn travelling clothes, the *seanchaí* favoured Mary with stories from the Blasket Islands, where she'd lived as a girl. His words mingled with the light and warmth of the fire until dawn, when the rooster reminded us all of the beginning of another day.

————

As I received them, the lessons were all equal in my mind, too. That they were presented in such a manner certainly played a part in that, but the fundamental reason was that life had heaped pain on me. I expected that it would continue to do so in new and novel ways, and I felt that there was no way of knowing which of the lessons I'd need most. So I soaked everything up like a sponge and was grateful for it all.

It was only as I got older that some of the lessons took on an extra significance. This could be for a number of reasons. Some taught me invaluable truths about myself, skills and ways of approaching the world that have allowed me to accomplish everything big I've ever accomplished. Others taught me how to view nature more fully, to see humanity's impact and to live in balance with the environment. Still others are just lovely memories, the kind of beautiful experiences that make life worth living—like lying in a soft bed with the sun on my ankles. And there are those that blend several of those reasons together.

———

I interacted with the physical remnants of the ancient Celts before I knew properly what they were. As I've said, their altars, ringforts, cairns and the like were everywhere in Lisheens. The locals told me that the valley was particularly thick with these ancient sites because it offered a view of both the mountains and the sea. These elevations had allowed for excellent tactical vantage points to watch out for various marauders and invaders over the centuries. An altar built in Lisheens offered communion with nature on its grandest scale while also ensuring that an attack wouldn't catch you unawares.

Even before my wardship, on visits to the valley I was brought on short pilgrimages to many of these sites. I would be introduced to the altar as though it were a person. These were formal proceedings with an air of ceremony, not casual greetings. My guide would first address me and then introduce me to the altar, beginning with the whole of the altar, and then highlighting certain features, helping me get to know my new acquaintance—*This is the place of cutting, this is the place of bleeding, this is the place that catches the moonlight, and finally there is memory in this altar.* It was on one of these forays that I discovered that my weeping stone was also an Ogham stone, an ancient site and a method of communication, something like a sacred newspaper or message board. Carved into the limestone were characters in the Ogham script, the ancient alphabet of the Celts. Though my stone had been special to me before I knew what it was, it was all the more so when we were finally properly introduced.

There was always a feeling of recognition in my bones when I first encountered these ruins. It was something like the sensation you get when your spouse, sibling or parent enters a room, before you've actually laid eyes on them—the hint of familiarity in the air before you consciously register their presence. Sometimes the feeling was reassuring and other times it was ever-so-slightly unsettling, the spookiness of unknown and invisible power humming in the air. On the sacred island at Gougane Barra, about ten miles north of Lisheens, the wall of prayers willed into existence by the monks who once lived in St. Finbarr's Oratory can make the hair on your arms stand up. When I'm there, I constantly feel as though someone or something is standing right behind me—not necessarily with malicious intentions, because its ancient origin was of divinity, prayer and personal sacrifice.

From the kitchen window of the farmhouse, I could see an altar to which I hadn't yet been introduced. Late one morning, in the soft stretch of time between the first round of chores and the start of dinner preparations, Nellie was combing her beautiful silver hair when all of a sudden, as though we were in mid-conversation, she began talking to me. "For the women of Ireland, back before Christianity laid its mantle over our lives, *Bealtaine* was a special day," she began. "And here it still is."

Bealtaine, she told me, was a celebration for women on the first day of May. That spring morning, the women of the valley would rise before dawn and gather in the field in front of the altar—she gestured out the window. They would collect dew from the clover on the ground and with it perform a ritual that was, she said, the secret to their beauty. Once the sun appeared, they would look to the altar itself.

From my seat I could see the altar had a long channel carved down its centre. "As the sun rises on the first day of the month of May, it centres its rays directly down that cut stone channel," Nellie said. "Out of the darkness of the night sky, a single shaft of light reaches for the altar. As it touches the stone, it looks as if the altar bursts into flame. It is a symbol of all the fires that have helped us in our passage through life's lonely walk."

In the ancient past, *Bealtaine* was the womb of the Celtic world. *Béal* means mouth and *tine*, fire. On the day, people celebrated all that was coming into life through the fire of the sun. The female principle—and all womanhood—was celebrated on that special day. The altar that was in full view from the window at every breakfast I enjoyed in Lisheens was going to take on a finer focal point for me.

I don't know if letters were exchanged between Uncle Pat and Nellie. I have my suspicions that a letter also made its way to Bona at St. Aloysius. In any case I found myself allowed to leave school and appear in Lisheens for the morning of May 1. The farmhouse was dark and creeping with shadows, the silence explosive with expectation. I was awake before I felt Nellie's hand on my arm. "Wake up, Gidl. The day is here." Then she led me out into the pastures barefoot, so she and I could feel the breathing of the ground underneath us. I remember the way the cold dew squished up between my toes as we walked.

Shamrock, *Trifolium dubium*, is a flowering clover that creeps among the grasses in pasturelands all over Ireland. It has been used as a fallow crop by Irish farmers since before the construction of Rome and its small yellow flowers are a favourite of the honeybee. "Just after the fire lands in the mouth of the altar," Nellie had told me, "the shamrocks begin to break out of their winter dormancy and push out their first green fingers of spring."

In front of the altar, we knelt in the grass to wait for the sun. All the other women of the valley were close around us. As we waited in silence, the group of us practised a controlled breathing meditation. With a gentle hand on my shoulder to get my attention, Nellie showed me the deep, slow rhythm of the breaths.

The first light of morning was little more than a pinprick over the lip of the altar's channel, but within moments the whole stone had come alive with light. It seemed to vibrate with energy. As the light intensified, I closed my eyes and waited for the feel of its warmth on my cheeks.

After a few moments, Nellie again touched my shoulder. I opened my eyes and saw that all around me, women were bending to wipe the morning dew from the fresh leaves of the shamrock. "Well now, Diana," Nellie said, "it's a fact that beauty gets noticed in this world, so let's not dawdle."

Wetting her hands on a plant, she showed me the pattern she used to apply the dew. She began running her fingers in sideswipes across her fine, high forehead just under the hairline, then moved down over her eyelids, gently patting dew into the corners of each eye. Next she worked her cheeks in a circular motion. Then it was on to her chin, which she massaged in upstrokes, and finally her neck, which received fifteen long strokes from her collarbone all the way up to her jawline. I mimicked her movements and then, faces wet, we walked smiling our way back to the farmhouse. We left the dew to air dry and for the rest of the day, my skin had a pleasant tight feeling as though it were being delicately pulled. This ritual, Nellie insisted, was what gave the women of Lisheens their beautiful complexions.

Later in life, I would confirm that *Trifolium dubium* does contain a biochemical for beauty. The upper area of the shamrock leaf exudes a pair of related flavonoids called hesperidin and hesperetin, most commonly found in citrus fruits. The morning dew holds these beneficial chemicals in solution and, when applied to the face, they encourage blood flow while gently tightening the skin's surface. Nature's anti-aging cream, so to speak. It improves peripheral circulation and has other advantages as well. By increasing oxygenation of the eye, it improves eyesight, particularly among people suffering from diabetes. And the collective controlled breathing is a form of meditation, which has been shown to decrease levels of the stress hormone cortisol.

At the time, though, I didn't need to fully understand *Bealtaine* to derive great happiness from it. The time with Nellie, the quiet communion with the other women, the relaxation of the wait, the fresh feeling of dew on my face, and the splendour of watching the sunrise set the hills and the altar aflame—all this was enough. That day has been fixed in my mind ever since as source of joy and female companionship.

The Field Experiment

THE IRISH SAY that laughter is a way to shorten your road. Well, Pat Lisheens knew the truth of that proverb.

He was an immensely loyal man who remained a bachelor because the woman he loved in his youth chose to marry another. He pointed her out to me at mass and told me her name in a whisper. She was lovely. He was a hard worker, who asked only that the land he toiled over day and night return the essentials we needed to live. And my God, was he funny.

His humour would change styles with the day, even the hour. He could be sharp-tongued and quick as lightning, tossing playfully barbed comments back at everything Nellie and I said or did. He had two languages to play with, Gaelic and English. He was a master of double-talk, that characteristically Irish habit of saying one thing but meaning another. Most of all, and most often, though, he revelled in playing the *amadán*, the fool, feigning stupidity to make us laugh. Sitting at the kitchen table, he might pick up a fork and say, "Ah, sure now, what's this for? What am I supposed to be doing with this?" and it was just funny. Something in his delivery had us falling off our chairs laughing at even the simplest, stupidest jokes.

One summer, I did all the transplanting with him and we laughed from one end of the field clear down to the other. He was never visibly worn down, regardless of how long he'd been at work in a day, and the mirth that twinkled in his eyes and played at the edges of his mouth made it a joy to labour along with him. He could approach each daunting task as though it were a bit of fun and so the road I walked with him was always shortened. With Pat at my side, I was willing to set out towards destinations I wasn't sure I could reach.

Out the front door of the farmhouse there was a long stone step that ran the full width of the house and the stables. This was a viewing platform from which to keep an eye on the fields. Standing on the step, the orchards with their apple, pear and Bing cherry trees were to my right, and the pastureland and grain fields sloped gently down in front of me into the heart of the valley, where the curlews sang their wistful cry. All of the fields were two or three acres in size, divided by hedgerows of gorse, ditches or low stone walls, and all of them had names. The *Gairdín* or "garden" field boasted the finest, greenest grass on the farm, and the animals were led there to graze before being milked. Past it was the hay field, surrounded by hazelnut trees, *Corylus*, which only grew about ten feet tall, bulking out instead as they aged, creating a dense living fence around the hay. To the right of the hay field was the *Droim*, named for the fact that it looked like the back of a massive animal that had settled down in the valley to rest. The *Droim* was planted with grain and the hedgerow there was rich with blackberries and raspberries.

In the last days of one of the first summers of my wardship, looking out from the step in the late afternoon, I spotted Pat at the edge of the *Droim*, scratching his head in consternation. I hurried down and came upon him standing in the dry August air studying the sheet of golden barley that covered the *Droim*, the heads of grain bobbing in the sun. Pat smiled a greeting and gave my upper arm a gentle squeeze, but there was an uncharacteristic twist of stress knotting his forehead. I asked him what was wrong. "Nothing that can't be managed, Gidl," he answered.

Pat walked along the edge of the field and picked a few heads of barley. He came back to me, rolling them in his big hand to shuck the grain as he walked. With a thumbnail, he cut across the seeds, holding out his palm as white flour spilled into it. The barley was ready to harvest. This was a good thing, I knew, and so I smiled my excitement up at him.

"Ah, not so fast, Gidl," Pat said with a slow shake of his head. "The tractor is broken in Kaelkill. It can't be fixed in time to make it here."

Pat didn't own a tractor of his own. When he needed one—to cut a field, say—he sent for its owner to bring his machine from another part of the valley. With that man's tractor out of commission, the barley would have to be cut by hand, and as soon as possible to avoid the crop over-ripening and spoiling. The *Droim* was a five-acre field, one of the biggest on the farm. Normally, facing such an emergency, Pat would call on his neighbours to help, but the entire valley was gathering in the grain—every farmer and farmhand was busy with their own work.

"I'll help you," I offered.

Pat's smile returned, wider than before, and his eyes lit with their usual mirth. "Aye, Gidl," he laughed, "you'll certainly have to." He patted me on the shoulder and again looked out over the field. "*Faoi mhaidin*," he said. "We'll start in the morning."

The dew was sparkling at our feet when we made our way back to the *Droim* the next day. Pat carried his scythe and sharpening stone. I had my two hands and a back that was willing and young. At the edge of the field, Pat made a first few cuts and then showed me how to fashion a cord out of the barley stalk itself. He demonstrated how to bind the sheaves and then arrange them in a pyramid-shaped stook that kept the heads off the ground, so that the barley grain could dry and cure in the air. I nodded my comprehension and, with a deep breath, we began.

Trailing Pat as he progressed with smooth, arcing strokes, I gathered the barley, bundling it together and arranging it in stooks. Pat was sure of himself and worked fast. As I struggled to keep pace, I worried about whether I would fall too far behind and let him down. By midday, with the sun high and hot over us, my back was screaming but my hands had become expert. I stayed bent, ignoring the ache that begged me to lie down or at least straighten, and made up ground on Pat when he paused to put the edge back on the scythe with his stone. We stopped only once, in the afternoon, to drink tea and gobble down a slice of buttered bread. When we heard the Angelus ringing out from the bell tower of the church in Kaelkill at six o'clock, there was just one small square of standing barley left in front of us. We'd finish before dusk. When the last stook was arranged, we made our way home through the gathering shadows. Pat could barely lift his arms to hang his scythe in the stables. In the kitchen, Nellie was waiting for us with a late dinner. We slumped, exhausted and triumphant into our chairs.

The "field experiment," as I've come to call that day, wasn't on the curriculum of my Celtic education, so to speak, but it taught me the most valuable lesson of my life. Looking out on the field that morning, I didn't believe we could finish the cut. It was too big a job and I was too small a helper. But because I loved Pat and loved the farm and the *Droim*, because I loved that crop of barley and couldn't stand to see it spoiled, I was prepared to try. I took a deep breath and a first step and both eventually led me to the knowledge that I was capable of things bigger than I'd imagined. I think every child should have an experience like the field experiment, something that thrusts them down a road they don't think they can ever reach the end of—for love of themselves and other people and the world. When they eventually come to their destination, they will realize, as I did, that even the impossible can be accomplished if you're willing to take the first step and give it everything you've got. And then, like me, they will know they can do anything.

Where Are the Trees?

I CAN'T REMEMBER exactly when I first realized that all the trees were missing from the landscape of Lisheens. I carelessly asked Nellie one day about Irish oaks. I told her that I had never seen one. She gave me a strange answer. There was an avenue of Irish oaks over in Glengarriff, she said, adding that Queen Victoria had planted them there at her hunting lodge. I thought about Nellie's answer and connected it to the fact that my father's family had forests, too, in Ireland, England, France and parts of Arizona and New Mexico. Rich and prominent people still had forests, but ordinary people did not? I never asked Nellie about where Ireland's forests had gone. As I look back, the answers were staring me in the face from the land, the people and the cruelty in the pages of history.

From my first proper introduction to trees with Dr. Barrett in the arboretum at Belgrave Place, I was enchanted by them. The trees I knew were the most wondrous and dependable beings in my life, and I was eager to meet more of them and to learn about each one. I looked for trees everywhere I went, but tended to come across them only if my parents dragged me along to visit acquaintances on an English estate. The scarcity of trees in the Irish landscape didn't strike me as odd or foreboding, though. Rather, I believe that in my early life I took it as a given—proof that trees were special—and failed to grasp that Ireland hadn't always been so treeless.

When my weeping stone was properly introduced and I learned its identity and purpose, I also got my first good look at the Ogham script. In ancient times, the Druids carved messages into the sides of these large rectangular stones using the alphabet created by a young Celtic man named Ogma. My weeping stone was covered with these thin characters assembled from parallel and intersecting lines. Though some had been rendered illegible by time and weather, many were easy to make out. Some days, when I had finished looking out at the hills and the sea, when nature had bled into me and calmed me down and the throb of unshed tears had passed, I would turn to study the stone itself, this ancient object that brought me so much comfort. At first, on my own, I could make nothing of the script, but after I worked up the courage to ask about the markings (this was before my lessons began), Nellie told me about the written language of the ancient Celts.

The Ogham alphabet comprises nineteen characters and most are named after trees. With her finger on the surface of the kitchen table, Nellie traced the shape of each as she named them for me, first in Gaelic, then in English. *Ailm*, pine. *Beith*, birch. *Coll*, hazel. *Dair*, oak—here she paused out of respect for the favourite tree of the Druids. *Eadha*, aspen. *Fearn*, alder. *Huath*, hawthorn. *Iúr*, yew. *Brobh*, rush. *Luis*, rowan. *Muin*, blackberry. *Nion*, ash. *Aiteann*, gorse. *Úll*, apple. *Ruis*, elder. *Saileach*, willow. *Cuileann*, holly. *Fraoch*, heather. *Straif*, blackthorn, the last of the bunch. (I've laid them all out—each letter, its tree or plant and their significance and uses—in part II of this book.)

The steady rhythm of Nellie's words and the movement of her fingertip against the wood of the table was like an incantation. I was so excited by the magic in the moment—the feeling that she might be summoning a forest from thin air—that I didn't wonder how Ogma had known the names of trees that didn't grow in Ireland. Instead of asking Nellie where the Druidic *ollúna*, or experts, had encountered the pines and oaks of their alphabet or asking where the aspens and alders of the ancient Celtic culture had gone, I just thanked her. Trees were so regal I took it as a given that people would use them as the building blocks of a language. It made good, sound sense.

But of course, the answer to the riddle of the trees was that Ireland had been deforested. After the Iron Age, deforestation began in earnest in the time of the Penal Laws, the five hundred years of English occupation that my mother had shown me a glimpse of in the story of the priest leaping the gap at the Pass of Keimaneigh. The Celts were a woodland people, their culture born from the deciduous rainforests that once covered much of the country. But as the English subjugated the Irish, they cut down these ancient woods. They cut them for the lumber needed in their naval yards and the charcoal to fuel industry. They cut them to empty the land of places like Lackavane, where the Irish could hide, regroup, plan and launch counterattacks. And they cut them to sever the most tangible link the Celts had to their culture and language.

During the penal times, the Irish weren't allowed to own trees or even certain seeds—they were effectively only allowed to grow potatoes for food. None of my teachers in Lisheens, most in their eighties and nineties, would have seen much in the way of trees in their long lifetimes. In the valley, holly and hazelnut grew in low, wide habits like bushes. The knowledge of plants passed on to me as a child, I later learned, was one that had been nearly destroyed long before my birth and gradually rebuilt. It contained only the smallest remnants of the Celtic wisdom about trees and their uses. How could it have been otherwise? There were no trees to learn from.

Or almost none. In all of the Valley of Lisheens, there was one tree still standing. And as luck would have it, this sole survivor of the great forests of Ireland stood in Nellie's farmyard.

It was a European ash, *Fraxinus excelsior*, and it must've been left alone to grow for hundreds and hundreds of years. It was an enormous tree. Its great canopy spread over the cowsheds and reached so far into the sky that its leaves cleared the barrier of the back hill and could be seen moving in the breezes from distant Bantry Bay. How they kept that ash I don't know; it didn't occur to me to ask as a child. I took its existence as a given. The house is there, the cows are there, the ground is the way it is, the tree is there. I wish I had thought to ask: Why is the tree there? And why this tree and no others?

I always thought of the ash as Nellie's—not her possession, but her ward. She would walk out to the dairy yard, stepping into the immense shadow of the tree, and fall into meditation. Watching Nellie and the tree from the window or the step, or from a small distance in the yard, it seemed to me that they spoke to one another, that there was a form of telepathy between them. Nellie would eventually open her eyes and shake her head free of her trance. Filled with fresh energy, she would brush some invisible speck of flour from her homespun skirt and look for me. She always looked for me.

Later, I gained a fuller understanding of the communion between Nellie and the ash. To the Druidic mind, trees are sentient beings. Far from being unique to the Celts, this idea was shared by many of the ancient civilizations that lived in the vast virgin wildwoods of the past.

The Celts believed a tree's presence could be felt more keenly at night or after a heavy rain, and that certain people were more attuned to trees and better able to perceive them. There is a special word for this recognition of sentience, *mothaitheacht*. It was described as a feeling in the upper chest of some kind of energy or sound passing through you. It's possible that *mothaitheacht* is an ancient expression of a concept that is relatively new to science: infrasound or "silent" sound. These are sounds pitched below the range of human hearing, which travel great distances by means of long, loping waves. They are produced by large animals, such as elephants, and by volcanoes. And these waves have been measured as they emanate from large trees. Children can sometimes hear these sounds, and I believe that Nellie could sense them and even to some degree interpret their meaning.

Nellie's ash was a *bile*, a sacred tree, and one from which Druidic physicians had cultivated many different medicines. Nellie retained some measure of that knowledge and after meditating, and catching sight of me, she would carefully collect any fallen twigs, gathering them in her skirt. In this way, the apron of the ash, the open space around the living tree, was always clean and neat as a new pin.

———

The summer I turned fifteen seemed to blow in a drastic change in the weather. Both June and July that year had been parched months, but shortly after my birthday in late July the rains came. There is a word in Irish, *báistí*, which describes an agricultural disaster in which heavy rain floods the fields, drowning crops and leaving the bogs too sodden to cut turf for fuel. That's what happened that summer. When the downpour finally broke after several days, Pat Lisheens was keen to get his work done in the bog. Turf had to be cut and dried to heat the farmhouse in the wet winter months.

The field experiment had affirmed the joy and satisfaction I got from helping Pat. When I found him early that morning, perched upon the cart, which was laden with sisal bags and a special turf-cutting shovel called a *sleán*, I eagerly climbed up beside him. The horse stirred. We set off down the laneway, bordered by holly on both sides. Past the old slate well, we steered across the fields, finally stopping at the wet rushes that bordered the long, narrow bog.

Turf is cut in rectangles like large bricks. As the cutter works across the bog, they leave a sheer cutting wall in their wake. In Pat's bog, this wall was more than eight feet deep, with murky, black-brown water carrying a rainbow slick of oil in the trench at its base. As Pat climbed down and eventually sank knee-deep, he told me that this was a particularly dangerous bog. "More than one animal has drowned in here, Gidl," he cautioned. "So mind yourself."

My job was to stand at the top of the cutting wall, receiving the chunks of turf Pat lofted from the blade of his *sleán*, turning them so they dripped out some of their load of water, and then forming a series of four-piece stooks so they could dry. I had to test each stook by hand to ensure that it was firmly planted and capable of withstanding the wind, and Pat would return to the bog later to rotate all the sods so each side caught the drying heat of the summer's sun. Pat's crop today was the denser sod, deeper down in the bog, which was valued for its slower burn in the frigid winter months. These kinds of turf sods behaved more like coal when burned.

We hadn't been working long when Pat's *sleán* thudded to a stop against something solid. First tracing its edges with the shovel, Pat dug the mysterious obstruction out and then picked up the dark lump and dropped it at my feet on the top of the wall. It was sopping wet and oozing a brown liquid, and it had a strange smell. Pat turned it over with the *sleán* and then looked up and broke into laughter at my confused expression.

"Well, Gidl," he said. "Run a guess at this one."

I got on my knees to examine the lump. I was hoping it would be a skull, but that didn't seem to be the case. "Not a skull," I said. "It's the wrong shape."

"That, there, is the heartwood of an ancient Irish oak," Pat said. "And I'm willing to bet that the tree it came from was growing when the court of Tara was built some two thousand years ago. This type of wood is called 'bog oak' and sculptors favour it. Gidl, you're looking at all that remains of the great forests of Ireland."

I had now learned about the English occupation and the loss of the trees, but here at my feet was the first proof. Proof not only that Ireland had once been forested—a confirmation that was amazing enough on its own—but that human beings were the reason the forests were gone. Nature's changes weren't limited to the shifting seasons, it seemed, and the landscape of Lisheens wasn't a fixed one, as I'd taken for granted. Trees and plants, the beings that most fascinated me, could just disappear. And, sickeningly, some people had worked towards exactly that outcome.

I was so upset, I sat down right where I was, picking up a long smear of brown along my leg from the heartwood. Pat, back at work in the trench, couldn't see my face as I looked around the bog and back up the valley towards the farmhouse and tried as hard as I could to summon the trees back into existence. I had tears in my eyes as I fought past the catch in my throat. "Pat," I was at last able to say. "Think of what it must have looked like here when there were forests."

His reply came over the lip of the cutting wall with the usual speed. "Ah, Gidl, it takes a bit more effort to imagine that from down here."

A Duty of Care

WHEN MY BREHON wardship began, the teachers of Lisheens did not even know how long it was going to last. They simply continued with the Celtic Triad, the nurturing of body, mind and soul. When they were all satisfied that "the Gidl, the *leanbh*, was coming along," they decided that I had absorbed the ancient knowledge I would need, including an aptitude for internal telepathy needed to protect myself as a woman. By the end of the third summer, my mourning had subsided to the point that I could laugh along at the best of Pat Lisheens's jokes. I could at last "walk alone."

Amazingly, I did feel fully ready to care for myself. On the cusp of sixteen, I was becoming independent. I had been running the household in Cork for two years, along with attending school. The Lisheens elders had given me the warmth of their love to stabilize my life and give me courage. Together with that, the judge, my barrister and my solicitor decided that I could live with Uncle Pat until I turned twenty-one. Though the curfew remained, for the most part I was able to push my great fear of ending up in the Magdalene Laundry at Sunday's Well into a corner of my mind, though it still came out to haunt me occasionally. Uncle Pat had learned that he could trust me. I always left a note on the kitchen table with the name and address of any friend I was going to visit. The respect that had grown between us seemed to deepen into a bond of love. The nuns at St. Aloysius gave me extra care; their thoughtfulness protected me, too. The art teachers at the Cork School of Art took me on as one of their more serious students. In addition to this, I'd learned how to design and make all of my own clothes. In short, I had my feet back under me and planted firmly on the ground.

What I hadn't expected, and wouldn't fully recognize until much later, was that I also felt ready to care for things beyond myself. After learning that I had value just as I was and without having to do anything to earn it, it was a small next step to see the same thing in other people. One small step more and it was clear that everything in the natural world possessed innate value and was owed the same duty of care I granted myself and the people I held dear. This belief, that a person should love others and nature as much as they loved themselves, was at the very heart of Celtic philosophy. It had been drilled into me with every lesson. I can say now, after years of looking through eyes tinted with gorse, heather and sea breeze, that I can imagine no more fulfilling and joyful way of seeing the world.

Of course, loving trees as people didn't exactly require the same faith as a leap over the Pass of Keimaneigh—trees were some of my oldest friends, after all. I suspect this is true to some degree for many people. It's not hard to love something so huge and magical. Take a child to the ancient redwood forests of western North America and tell me their reaction isn't a loving one. What doesn't come as naturally is the understanding I first gained in the turf bog: that human action can drastically impact the natural world, and that, as a result, each of us bears a responsibility of care for everything around us.

The Celts built this responsibility into their culture. On that first medicine walk with Nellie, she taught me that the most fundamental rule in harvesting nature's bounty was to "always leave enough for the seventh generation." It's my belief (cultivated later, in the stacks of the University College Cork medical library) that this warning was learned by the Celts and many other ancient peoples because of a crisis in Greece. In the time of Christ, there was a species of *Umbelliferae*—the family of celery, parsley and carrots—that grew in very select parts of that Mediterranean region. This unusual plant, a giant fennel, was used as a contraceptive, likely drunk in a liquid decoction. Through overexploitation, the species went extinct, sparking a population boom that put tremendous stress on agricultural production and led to food shortages and other deprivations. To ensure that something similar didn't happen again, an oral warning was needed. The "seventh generation" guideline is just such a warning—a reminder that future generations will follow and that they will also need what's provided by the natural world. It is a caution against the greed and unnecessary accumulation of goods that define and drive so much of our modern society.

———

At the end of that third summer, Nellie walked with me to Nurse Creedon's surgery, near the church in Kaelkill. When we arrived, I found the place stuffed to the rafters with all of the people of the valley who had contributed to my education over the course of my wardship. This was my graduation.

I was guided to the centre of the room. With all of those dear old faces around me, I felt cocooned in love, immersed in a warmth like nothing

I'd ever experienced. I felt like I could accomplish anything. I would need that belief.

Mary Cronin was there to finalize my wardship by looking into my future. Mary was the prophet for the area; second sight was passed down her family line. I had always been afraid to have Mary look into my future, terrified at the ugly things she might see, given what I'd already experienced. Biddy had died in her sleep just months before. I did not want to hear of any more deaths. Here, though, surrounded by my teachers, I was ready.

Mary told me many things. She told me I would marry a man who would guard me like a lion; I can confirm that my husband, Christian, is certainly that. She described the home in Ontario that Christian and I have now shared for nearly half a century—a house "in a land with lakes and evergreen trees." She told me I would "climb the ladder of success" and know it when I reached the top, and that I would take up writing in middle age, which I did. And she described many of the most wonderful and inexplicable experiences of my life, some of them connected with the First Nations, telling me I "was the woman protected by the Golden Eagle." Most everything she painted for me has come true, though, having lived a very simple life, I'm still made uncomfortable by the idea that I might be "successful."

These predictions were Mary's visions, coming from Mary alone through whatever her second sight gave her. But as I felt her words were nearing an end, she extended her arms to the room and I knew she was indicating that the last of what she had to say came from everyone, from the whole of the valley and the Celtic tradition.

"Diana, you have been given a sacred trust," she said, her voice breaking with emotion. "We are old and can't live forever. When we are gone, you will be the last voice of the ancient world of Ireland. There will be no more after you."

I shouldn't worry at the weight of my responsibility, Mary assured me. I was descended from the kings of Munster; my grandfather was the Brehon judge Daniel O'Donoghue. I had it in my blood to hold and safeguard the Celtic knowledge my family had protected, just as that knowledge would safeguard me. But I had a mission.

"You've got to bring this information to the New World, Diana," she said. "The youth of Ireland are deaf to it and blinded by the things they want. Across the ocean, people will one day crave the ancient knowledge and recognize it as the only way to save themselves and the world.

"You just have to wait for them to be ready. You'll know when they are."

The Science of Ancient Knowledge

I TUCKED MARY Cronin's words away in the vault of my mind, a tantalizing but still abstract prophecy. Only sixteen at the close of that final summer of my wardship, I wasn't yet free of the judge and lawyers in Cork, let alone free to leave Ireland in pursuit of my life's true purpose.

I was a different person from the girl I'd been when I was orphaned—I now had the protection and fortification of the ancient knowledge of Lisheens. But while the rest of my life had also improved, I was still terribly lonely, and the shift from the warmth and belonging I felt in the valley back to my mostly solitary existence in the city at the end of that summer was an awful one. Now, though, when I found myself thinking, *Why did all this have to happen to me?* I had an answer. Mary Cronin and the women of Lisheens had told me there was a reason for it. Having been selected for such an important task, I realized I had to deal with whatever hardship came arm-in-arm with it. So I began to take the loneliness and estrangement I felt from my classmates more as a matter of course, or maybe as the burden that came with the incredible gift I'd been given. There was real comfort in that acceptance; it gave me permission to show my true self. I couldn't control how anyone else felt about me, so why should I hide my intelligence or any other part of me for fear of making people uncomfortable? Math was poetry to me; books were food and science was air. It no longer mattered to me who knew that.

I continued to be taught math at the breakneck pace I loved by the professor brought in by the headmistress, moving on to university-level courses while still at St. Aloysius. Those classes remained a kind of open secret. None of my classmates or regular teachers ever spoke to me about them, and the top marks I earned in all subjects went similarly unremarked upon. Uncle Pat left my report cards unopened on the table, along with the bills for the water, heat and electricity, until they finally made their way, still unopened, to the bin.

Despite never being directly addressed—at least, not to me—my growing academic reputation at last led to a social life outside Uncle Pat's library, even if it was a transactional one at first. Girls started to appear in the school's cloakroom, the quiet place where I liked to settle in to read, looking for help with the subjects that were causing them difficulty. I offered guidance on essays in English and Gaelic, along with instruction in mathematical or scientific concepts. I was the tutor tucked behind a veil of heavy coats, hats and scarves. "Find Diana, she's in the cloakroom. She'll help you."

Though Uncle Pat didn't care to know my marks, he remained deeply interested in what I was actually learning. Adding to the ideas and images we dug up and shared from among the treasures in his library, I brought in fresh discoveries from the outside world. It was a great joy to see something that had seized my imagination create a similar effect in his, and that exchange—the basis of our relationship, really—contin-ued to deepen our connection. I also started to earn money by taking in dressmaking work and designing and producing posters for things like the local dances. I used that extra income to buy better food for us, though that didn't seem to register much with Uncle Pat—not the way a new idea did, at least.

So I spent my days at school and tending to the house at Belgrave Place. Uncle Pat and I often shared dinner and some time reading in each other's company, and most of my other spare moments went to art. I carried my paints with me to Lisheens in the five or six more summers I visited, after my wardship and before I left Ireland. Whenever Pat Lisheens didn't need my help, I'd go down to the Owvane River or ride my bike to Ballylickey with my paints and brushes, a jam jar for water and some rags, and I'd sit down in the grass and paint all day. The whole of Nellie's house was full of my work. I submitted some pieces for competition when I was fifteen or sixteen and won a scholarship to the Slade School of Fine Art in London, but I worried I'd starve if I became an artist, so I didn't go. I did, however, keep up with my lessons at the Cork School of Art. Every day after the final bell rang at St. Aloysius, I headed there and worked under the instruction of Mr. Teigen until the moment I had to rush home to make curfew.

Painting taught me a way of looking at the world, and at nature in particular. It taught me to soak up the beauty around me and also to register the finest details. I discovered so much about the way different leaves are constructed by trying to recreate them on paper, and the same held true for whole trees and grasses and anything else my eye landed on—perhaps something as simple as the stem of an apple in a bowl on Nellie's table. Making art also stoked my creativity, something Uncle Pat encouraged as well. He may not have known that a girl needed to eat, as he paid so little attention to such things himself, but he raised me to be intellectually independent, to come to my own conclusions and resist the strictures with which other people tried to constrain my thinking. Albert Einstein once said that the greatest gift a scientist can have is imagination, and he was absolutely right. We can shake free of others' demands and expectations, but we are always bound by the limits of our creative powers. You can't dive off into new intellectual territory if you don't first imagine it into existence. That gift to dream, to go beyond what seems possible, was given to me through art and in Uncle Pat's house.

By the time I graduated from St. Aloysius, I had made a couple of close friendships, but I still wasn't invited to the dances and parties that had come along with the arrival of boys in the lives of my classmates. My name continued to set me apart, as did my financial situation at home. Unlike many of the girls at St. Aloysius, Uncle Pat and I didn't, for example, have a country house by the seaside in Crosshaven. My intelligence also separated me, even if it also made me an object of some curiosity.

With high school completed and the realities of adult life lying ahead of me, though, the sense of possibility that my intelligence lent me—and the real possibilities it opened up—started to attract more people to me. I remember, in particular, a group of six girls that I started to know around that time. They weren't close friends, but they invited me every Tuesday evening for tea and cakes. None of them was university-bound. Instead, not yet twenty, they were entering the working world.

I was headed for University College Cork, and they looked up to me because of that. But they also liked me because I wasn't stuck up about that academic future. I knew that continuing my education didn't make me better than them, and they appreciated my lack of airs. Those women were the first social group of people my own age I'd ever belonged to, and the fact that they didn't just put up with me, that they actually wanted me there among them, reassured me that the choice to let my true self shine, consequences be damned, had been the right one. Within the next couple of years, all of them were married and starting to have children. I continued to visit when I could, but I had other plans for myself, which was one of the things they appreciated most about me. I had my eye on becoming the chair of a biochemistry department.

From a very early age—thirteen, maybe—I made a firm rule for myself. Never waste a minute. The most precious thing you have is your time. Our lives are narrowed by birth and death and in between lies everything you set yourself to achieve. We only have a counted number of minutes given to us, and my parents' deaths had shown me in the most painful way how unexpected the cut-off could be. So, faced with the choice of what to study at university, I decided to satisfy my thirst for knowledge as completely as I possibly could.

I took a double-first in medical biochemistry and botany, working towards full undergraduate degrees in both fields simultaneously. In my first year, I still hadn't received a penny of my inheritance and remained reliant on the courts to cover basic expenses. So I got a job as a teacher's assistant, a "demonstratorship," as it was called, that made me responsible for setting up the botany labs and maintaining the school's botanical collection. In place of the photographs that are used now, botanical drawings were still part of the curriculum, and I got another job teaching fellow botany students to draw—learning and instructing at the same time. When classes started, I remember feeling as though I'd found the taps that fed my brain and spun them wide open, with joy at the saturating rush of information.

There were wonderful experiences in store. My second lab in botany was on *Chondrus crispus*, the seaweed known by the common name

Irish Moss. Seeing it there on the lab table was like running into a dear old friend in an unexpected new place. My great-aunt Nellie had taught me about that seaweed in Lisheens. Growing in the mid-tidal zones, it looks from a distance like a big double- or triple-flowering peony, mahogany or blood red in colour and surprisingly bright. It extends up from a holdfast, a stem with which it fastens itself to the rock. Aunt Nellie had taught me that in the time of the Great Famine, more than a hundred years ago, people were prone to tuberculosis stemming from malnutrition. *Chondrus crispus*, she said, held the cure to that ailment. Flipping the plant back from its holdfast, you removed it from the rock, took it home and boiled the whole thing, causing it to release a gel-like mucilage with powerful healing properties, effective in the treatment of tuberculosis and also good for your intestines.

Dissecting it, I found that there was indeed a sugar mucilage form in the structure of the plant. When the lab let out, I hurried to the medical library and hit the stacks. With books and anthologies of the journal *Botany* piled around my seat, I discovered that the jelly derived from *Chondrus crispus* has strong antibiotic properties, as well as the ability to remove radioactive strontium from the body.

The feeling this confirmation of Nellie's teaching gave me is hard to describe. I loved my teachers in Lisheens, but I hadn't completely ruled out the idea that the things I'd been taught there were just old superstitions. I needed to confirm them for myself. There was always the chance that there would turn out to be nothing of import in the plants they'd emphasized to me, and nothing more to the ancient knowledge than beautiful clouds of vapour. Reading in a book in the university medical library that *Chondrus crispus* actually contained medicine that could do what Nellie had said it would, after drawing the jelly she'd told me it contained out of the plant with my own hands, was my first irrefutable proof that the lessons of my wardship were based in something real. I felt relief, accomplishment and the happiness that comes from coming face to face with the truth of nature. And I felt something else, too.

The knowledge I'd received in Lisheens came in the oral form and, apart from the Brehon Laws themselves, it existed in no other form. But there in the medical library, I was looking at the exact same knowledge, derived and presented in a completely different way—written down in a book. In that moment, I saw that I could serve as a bridge between those two worlds, the ancient and the scientific. The realization was hugely motivating. It made me want to test the validity of everything I'd been taught in Lisheens immediately, to put the knowledge on trial at once.

As energized as I felt, not everything could be confirmed as quickly as the medicines of *Chondrus crispus*. My Celtic education wasn't one of instant knowledge, but rather of truths gleaned from thousands of years of continuous observation and experimentation. Sometimes significant time passed between the start of my investigation and the point where I got confirmation; I would verify bits of what I'd been taught but only assemble the full picture years later. I didn't fully grasp that in my first year of academic study, but it didn't dampen my enthusiasm. After all, I'd been conditioned in Lisheens to see that even the smallest piece of knowledge held as much value as the whole picture. I dove into the trial process immediately and wholeheartedly.

I always started with the plant itself. I would examine it first with my artist's eye, searching out and noting the crucial and unique features that would infuse its truth into a botanical drawing. Next I would dissect the plant, breaking it down to the smallest component parts and examining each under the microscope. Basically, I would absorb every bit of information I could through my own five senses, and from there I would develop a baseline understanding of how the plant was built and how it behaved when you acted on it in certain ways. Armed with that first-hand knowledge, I would then head to the stacks to discover what knowledge others had to contribute to what I'd picked up with my two hands, my own two eyes and my brain.

The ancient knowledge often worked as a signpost or guiding star in this trial process. My teachers in the valley might have indicated that a particular plant was good for poor circulation, which I'd take to mean heart trouble. I would then know to keep a particular eye out for the presence of any chemical known to benefit the heart. "Well, Diana," they might have begun, while cradling a small, five-pointed, yellow flower in the crook of two fingers. "St. John's Wort, as you see here, has a strong medicine for nervousness and mental problems." I would later find out that St. John's Wort contains phytochemicals such as hyperforin, which increase the effectiveness of dopamine and serotonin in the brain. The plant is as effective as many prescription anti-depressants, and may in some cases be more effective.

My twin degree programs armed me with the ideal mix of disciplines for testing the knowledge of Lisheens. The combined effect of the ancient knowledge and my university studies allowed me to see links between the medical world and the botanical world early on. Soon, my seat in the library could have had a small gold plaque on it: *Reserved for Ms. D. Beresford.* That seat was by a window that looked out over the university quad, and everyone respected it and left it free for me. I could look up from my investigations and see groups of students walking around and gasbagging below me. The sight always filled me with love, an undefined love for other people and a wish for their happiness that radiated from my heart and seemed to settle, in particular, in my cheeks. That undifferentiated love for humankind is still one of the great motivating factors in my life and work.

They may not have phrased it this way, but the people of Lisheens knew that there were links between classical botany and human biochemistry—between the natural world and our health. After I recognized just such a link in the medicinal properties of *Chondrus crispus*, I started to look at the biochemistry of other plants from my seat in the library. I also got interested in pharmacognosy, the study of medicinal drugs derived from plants and other natural sources, and the no-doubt more familiar pharmacy, the science of preparing and dispensing drugs.

Looking at the biochemistry of a plant I'd observed and dissected, I might find alkaloids, fats, sugars, lipids. And as my understanding of human biochemistry grew, I began to grasp the effects that each of these secrets hidden within a plant could enact on the body. Your body needs twenty-two essential amino acids. They're built into proteins and you have to eat them—if you don't have them, you've got a problem. There are three essential lipids, also called essential fatty acids, going into your nervous system, and if you don't have *them*, you've got a problem. And then there are all sorts of sugars, which can be held in different polymeric forms, and the microelements—like sodium, selenium, and potassium—that are themselves crucial to your body's proper functioning. The plant kingdom supplies all of that for your body, a relationship that is still fascinating to me some fifty-five years after I first encountered it.

The rough model of investigation I invented for myself as an undergraduate is the one I've used for my entire scientific career. The sources of knowledge it is based on and employs—the ancient Celtic teachings, classical botany and medical biochemistry—shaped my thinking in foundational ways. When I investigate a plant, my mind tends to work in two directions at once. From my understanding of the plant, I work towards the human body; and from my understanding of the human body, I work towards the plant. I have never failed to find a point, or points, where they meet. Every plant is intimately tied to human beings and our health. The people of Lisheens knew this and a great many other things. From those first investigations through to the present day, I have been able to scientifically prove almost everything they taught me during my wardship. The only thing that has eluded my understanding is telepathy, the invisible links that they taught me exist between human minds. That one I'm still working on.

———

My undergrad degrees were both three-year programs. We studied from September through mid-June, when we were given six weeks off. I went to Lisheens during that time away from school, but exams were held the moment we returned in August, so I had to bring my books with me. There was no electricity in the farmhouse, and Pat and Nellie knew I needed light to read by. I would study physics by the light of a paraffin lamp they bought especially for that purpose.

In the summers between the end of my wardship and the year I left Ireland, I continued to visit my Celtic teachers. Having grown to love them for the care and attention they showed me, I'd stop in for tea, to help with this or that around the house or just to talk. They had all been old women and men even as they were instructing me, and in that time they started to die. Each of these losses hit me hard, but now I can marvel at the fact that my wardship happened at all. My time of intense need came just as I was old enough to understand what they taught me. It came when the larger dynamics of the valley—and Ireland and the world—had devalued the ancient knowledge and left no one else around who cared enough to learn it. And it came right at the wire, in the final years that there was still anyone alive left to teach me. You may not believe in fate, but certainly you can see why I feel that safeguarding and sharing the ancient knowledge of Lisheens is my duty to perform.

———

Just days into the last year of my undergrad, my botany professor, Dr. Oliver Roberts, had a heart attack. He survived, thankfully, but was told by his doctors to lay low until he regained his strength. He sent for me from his sickbed and when I arrived, he said, "Diana, I want you to finish my honours lectures." He had only given the first two.

Taking on this task meant that I would be teaching the university's third-year botany course while I was myself a third-year botany student. I said yes on the spot. Roberts meant only for me to give the lectures he had written, but I felt that in order to teach botany I had to understand the entirety of the plant world. In prodding me towards a view of plants wider than any I'd previously attempted to take in, Roberts gave me an incredible gift. I grew to look at the world as a single unit, to see the entirety of the global garden and the connections between all living species for the first time—and also to decipher the potentially disastrous consequences of climate change.

I approached my new role as lecturer as a larger version of the pursuits Uncle Pat and I had set out on. Tracking the evolution of plant life from chlorella, the aquatic unicells with which photosynthesis began, to algae, then multicellular algae, fungi, mosses, ferns, evergreens and finally the angiosperms—the flowering plants and trees that are as biologically complex as we are—I became fascinated with the question of what came between these major steps and fundamental categories. Life couldn't simply leap from a fern to an evergreen; there had to be something lost in the margin between them. I decided that the students—and I was one of them—needed to go back in time, to speculate on the species that forged the connection between the ferns and the evergreens, so I added to the lectures. Focusing on those species—like the tree ferns, *Dicksonia antarctica*, and the intermediate species of the strange *Welwitschia* family, and palm trees in the *Cycadaceae* family that are found on tiny islands in the South Pacific—is what first got me thinking about climate change.

Earth's atmosphere at the time of change from the ferns to the evergreens had concentrations of carbon dioxide too high to sustain human life. Fortunately for us, humans didn't yet exist. If we had, we would have suffocated. Over the next 300 million years, the ferns, then cycads, then long-lost extinct evergreens and then gymnosperms and finally the flowering trees oxygenated our atmosphere. Green molecular machines continued to evolve, converting carbon into stalks, trunks, leaves, flowers and breathable air, each more powerful than the version that preceded it. Trees don't simply maintain the conditions necessary for human and most animal life on Earth; trees created those conditions through the community of forests. Trees paved the way for the human family. The debt we owe them is too big to ever repay.

That process struck me as very important, so as I redid Roberts's lectures, I inserted it into them. And from there it was a short hop to my first understanding of the potential impact of human actions on the environment, both for the planet and for our own health. The truth was right there, so simple a child could grasp it. Trees were responsible for the most basic necessity of life, the air we breathe. Forests were being cut down across the globe at breathtaking rates—quite literally breath-taking. In destroying them we were destroying our own life-support system. Cutting down trees was a suicidal act.

———

Roberts was well enough by the end of the year to mark our final exams, which covered everything we'd learned during the whole course of our studies. As I sat down to review what I'd learned in undergrad, I reflected on the other ways my time at University College Cork had changed me.

My physics professor in first year was John J. McHenry, who'd been a student of Wilhelm Röntgen, the man who discovered X-rays and went on to win the first Nobel Prize in physics. I would stay and ask McHenry questions after our lectures, and we became friends. There were about 150 students in the class and many would come up to me after I'd finished talking with McHenry and ask whether I understood what we'd been taught. When I told them I did, they asked me to explain it. I believe that there is an honour system, almost chivalric, when it comes to knowledge. What you learn, you share. I soon had almost an informal class that ran after most of my lectures—not just physics. Around the university, I was called "the brain."

This wasn't a replay of being the tutor in the high-school cloakroom, though. Finally seeing some value in myself and what I could contribute to the world, I'd cracked out of my shell with gusto. I was heavily involved in sports, primarily tennis, waterskiing and camogie, and I'd made some very big friends lifting weights with the rugby team. I acted in university theatrical productions. Instead of scaring people off, my last name now got me invited to hunt balls. I attended wearing dresses I'd sewn myself from Vogue patterns, and looked as though I had lots of money.

I didn't, of course, but I did eventually receive a first bit of money from the courts—three hundred pounds, which I used to buy a second-hand Mini. With that car a new level of freedom opened up to me. I remember one evening driving down to Kinsale with a bunch from the rugby team. I was pulled over and all the rugby fellows piled out of the car. The Garda looked at them and told me, "If you can get *that* crowd back in *that* car, I won't fine you." They tumbled back in and we set off again without a ticket.

In Ireland, final marks are announced at a public ceremony after the exam period. The roll is read in alphabetical order and I got to the ceremony for my final year a little late, arriving on my bike just as the *A*s were wrapping up. I listened past the *Be* names to the *Bu* names and on into the *C*s. I wasn't called. *Holy shit*, I thought, *I've failed this bloody thing.*

With the *D*s beginning, I got back on my bike and raced away to the row of shops on St. Patrick's Street. In shock, I peddled with reckless abandon. When I was out of breath, I stopped and went into Roches Stores' food section and bought a yogourt, which was a new arrival in Ireland at the time. I leaned against the counter and prepared to find some small consolation in my favourite treat. A schoolmate named Anne O'Leary walked around the other end of the counter and called to me, "Congratulations, Diana!" I was shocked by her cruelty and told her to leave off of me. "I've failed the exams," I said. "I wasn't even mentioned."

She looked at me like I'd grown a second head. "Well, that's because you came first."

I left the shop, put my yogourt in the front basket of my bike and peddled back. Anne was right; I'd finished with the top marks in the university. They'd announced my name before beginning the rest of the roll and I'd missed it. I was stunned: Anne was right. That I now had friends and even a measure of popularity didn't mean I was immune to my old insecurities, it seems.

The top marks I'd earned opened up several possibilities for me, but there were really just two frontrunners: completing a medical degree, the next logical step forward from my biochemistry degree, or taking my master's degree. As I mulled that choice over, there was one last thing I wanted to accomplish before I moved on from being a University College Cork undergrad. The botany department had a specimen collection—examples of all the species we were studying, preserved in formaldehyde. The collection allowed students to observe species first-hand and even enjoy a certain level of interaction with them, which had left the specimens pretty battered. I decided I would redo the collection.

I asked some of the best students in the department to help me. Ten or fifteen of us went down to the seaside, to a place called Glendor, where we stayed in a hotel for a few days while searching out specimens. I knew it was an area rich in rare plants. We broke into groups, each hunting for different things, and collected everything I needed to redo the plant collection.

On that trip, I noticed that rare species tended to occur in places where a river ran into the sea and fresh water and salt water mingled.

Those places, where you could find things like *Rhodophyta*, a beautiful articulated red algae, were also where fish and whales congregated; they were places teeming with life. I developed a hypothesis: that there must be an essential mineral carried from the land into the sea on the freshwater currents that created a platform for these rarities to flourish. Fifty years later, while filming my documentary, *Call of the Forest: The Forgotten Wisdom of Trees* in November 2015, the Japanese marine chemist Dr. Katsuhiko Matsunaga confirmed my suspicions.

Matsunaga and I were sitting on a beach at the edge of Ise Bay near Nagoya, Japan, when I told him of my early theory that the effect was the result of an essential mineral being chelated in the water. With a look of surprise, he told me that I was correct. Over decades of experimentation, Matsunaga and his team at Hokkaido University had proven that just such an effect existed.

As leaves decompose on the forest floor, they release fulvic acid, a humic acid capable of bonding with the iron in the soil. Flushed from the forest into the rivers and carried to the ocean, an iron-poor environment, this oxygenated iron boosts the growth of phytoplankton, creating a marine buffet. Prior to Matsunaga's discovery, a crucial hint was contained in an old Japanese saying, "If you want to catch a fish, plant a tree." The phrase was a signpost like the ones given to me in Lisheens.

There is, of course, a joy to finding confirmation of a suspicion you've carried around for half a century. But that satisfaction was tempered by the fact that Matsunaga's research was sparked by more than just the desire to test an adage. He and his team hoped to explain the mystery behind the widespread collapse of marine ecosystems along the Japanese coast. They showed that clear-cutting on the island nation was what caused the devastation. Cutting so many trees led to a decline in the amount of fulvic acid, created by decomposing leaf fall in the forest, leaching into the ground water that flows into the sea. This, in turn, reduced the quantity of iron in the island's coastal waters. The lack of iron stopped the division and multiplication of microscopic marine life, which meant a famine for the sea creatures that depended on that life to survive.

Cutting down trees, then, is not exclusively a suicidal act. It is homicidal as well.

The Sumac Flower

AS CHILDREN THE questions we're willing to ask are boundless. Why is the sky blue? What is life? What happens when we die? Where do babies come from? Why can't I stay up as late as you? Some of the answers are easily supplied; others have eluded human understanding for thousands of years. To our young selves, though, there is no difference between the small questions and the big ones. We follow our curiosity to the edge of our understanding and then ask whoever is around what lies beyond it. I have spent my adult life fighting to keep what I possessed as a child: the ability to see the biggest questions sitting inside the smallest ones and the willingness to try to answer them.

Looking back, I see now how lucky I was in my education, particularly through high school, access to my uncle Pat's immense library, and my undergraduate studies. So much of the academic world I have encountered since I left Ireland has seemed purpose-built to discourage big questions and to funnel knowledge down narrow, separate channels. Of course, I faced skepticism and roadblocks to my learning, but I was also supported by so many people—from Uncle Pat onward—who were happy and even proud to help me pursue the answers to whatever questions were driving me. I wish everyone was so lucky.

After I finished my double-first, there were questions pulling me towards medicine, but ultimately I chose a master's degree that continued my studies in biochemistry and botany. It was the path that allowed me the widest possible view and understanding of the natural world.

My subject was the hormones that regulate plants and the frost resistance of all species. I began with a question I felt no one had properly asked before, let alone answered: What are the margins of life? I wanted to understand the limits of the plant world, one of its keys being understanding the regulatory actions of hormones in plants. There was a beautiful sumac on the University College Cork campus that I visited whenever I had the chance. Its clustered, blood-red flowers appeared every fall and lasted all winter. I wanted the answer to a question big enough even for a child: Why did the sumac flower? And more specifically, what was it in the tree that caused that wonderful clutch of red blossoms?

At the time, at night after dinner, Uncle Pat and I were bandying about another important issue: the rise of global temperatures. This worried me a lot. All in all, my medical biochemistry and classical botany formed an arch of knowledge for me. Under that arch, I had also placed physics and chemistry. The arch became my viewfinder on the planet as a whole. I became intrigued by the pattern language I saw in nature. It looked to me that the regulatory patterns of DNA were similar in both plant and animal kingdoms. That idea shocked me—that the biochemistry of humankind was connected to the biochemistry of trees and plants, which could be seen in the hormones that regulated both plant and animal worlds.

Then, by absolute chance, while tutoring a medical student, I came across the photosynthetic reaction written out in what is known in chemistry as an empirical formula. I stared at in fascination, transfixed by the small idea I had in my head that would have such immense importance to my understanding of climate change. The photosynthetic reaction is the reverse of ordinary breathing. This means that plants and humans are connected by chemistry, by oxygen and carbon dioxide. These two molecules conduct both life and death. This, too, is an equation. What would happen if plants—let's say, forests—were removed from the planet? The answer is obvious. Life would be extinguished; death would come by heat or the greenhouse effect or insufficient oxygen. I began my master's in a rush.

I was given free rein over the university greenhouses, long structures with double-pitched roofs built on a low, flat stretch of campus. The theoretical parameters of my subject were truly vast, encompassing every plant species on Earth. The physical parameters of the greenhouses and my equipment, however, offered some definite limits. For my experiments, I selected species I knew would fit in the growth cabinets.

In effect, I was running a climate change study. Measuring the size, growth and ratios of a wide range of species under different environmental conditions, I investigated in as broad a sense as I could manage how plants reacted to changes in their environment—the impact of drought and cold on the organisms we need to sustain life.

This work yielded a great many discoveries. Every plant has a wilt point; I know that a pea plant, for instance, can survive to minus nine degrees Celsius, but it will not survive minus ten degrees. The concept of wilt points was known, but I discovered the exact wilt points of many species, which were unknown, especially in food crops. In species with greater drought and frost resistances, I identified the key role played by gibberellin, a plant growth hormone roughly equivalent to our human androgen hormone. A plant that has gibberellin is a hardy plant, and I discovered that this hardiness could be hybridized and passed along into more fragile species—that it was possible to induce vigour through crossbreeding. I discovered that, in trees, resistance to extreme climates is signalled by a blue-green tint in the leaves or needles. This tint is created by the reflection of sunlight off a cuticular layer that covers the leaves like a skin and is much thicker in cold-loving species. This layer also helps the tree retain moisture. I also noticed that ancient species of grain tend to have an abundance of lateral branches, known as tillers, at the base of their main stalks. This tillering allows the grain to grow successfully in high winds, a root and stem adaptation that will prove more and more critical as global climate change increases the frequency and intensity of extreme weather systems and events. Cellular colour, too, was an important adaptation to heat and cold. If you have a white version of a carrot, for instance, it can withstand less heat and cold than a red version.

In the years since my master's research, which I completed in 1965 at the age of twenty-one, these varying characteristics have become more significant to me—and to the future health of the planet. A few years ago, I flew to the coastal forests of western North America to select trees to be saved in a bank of genetic material. Many things went into identifying the best candidates, but among the chief qualities I looked for was the blue-green tint—exhibited in that case by the sequoias. These were the truly ancient redwoods, *Sequoia sempervirens*, iconic giants that are also fire resistant. It is my great hope that humanity will realize the need for concrete measures to address climate change before the global temperature rises much higher and the shifting conditions we have wrought render survival impossible for all but the hardiest species. Regardless of when we finally take collective action, though, the characteristics I identified in my master's research will serve as a guide to the species best equipped to survive in a changing world. I have followed that guide myself in selecting and saving ancient heritage species of rare plants and trees on the farm where I now live (more on the farm a little later). It was truly important work.

At the time, though, I couldn't see that. Or more accurately, I couldn't see that the work would be of interest to anyone but me. In the greenhouses, I had a couple of techs assigned to help with the counting and measuring, but mostly I worked alone. And though I was now an outwardly social creature—an extrovert, even—I was afraid to get too attached to any image of myself as free-thinking, competent and valuable. I wasn't conscious of that fear at the time because it stemmed from something that I tried my best never to think about: the threat of being sent to Sunday's Well. The terror of that place was too much to look at head-on, so I only properly reckoned with it in nightmares, where I was rendered utterly helpless, stripped of every freedom and choice.

Maybe a year after my parents' death I had climbed on my bike and ridden up to that laundry. Crouching outside the perimeter fence, I had spied on the priests as they came and went through the building's front door. As a little girl with no one on Earth to defend you, you become hyper-aware of body language; your sensory system is like a tuning fork that hums from head to toe in the presence of a threat. The priests wore long cassocks that nearly touched the cobbles. As they approached the front steps at Sunday's Well, each man would raise his hem clear of his feet with the same arrogant tilt of the wrist. In that movement, I smelled harm. Those men smelled of danger to me.

Even though I did my best to focus my thoughts elsewhere, keeping myself busy from morning to midnight, my terror at the looming threat of incarceration—there's really no other word for it—was always present in my mind. Avoiding it was exhausting; carrying it around was exhausting; diving into it was unthinkable. When the courts finally gave me agency over my own life and removed the threat of Sunday's Well, which happened the same year I finished my master's, I felt a great rushing relief, but that torrent also began to bring the fear and pain I'd stuffed down in my chest for so long to the surface. Trauma leaves a long shadow that, in my experience, never goes away and requires constant courage to withstand.

Unwilling, or maybe just not quite ready, to face all of that pain, I didn't stop to work through what had happened to me, to try to make sense of the damage done by my mother's coldness, my father's absence, their sudden loss, my grief and loneliness, Uncle Pat's initial negligence and the anxiety of dealing with the courts. When I reached my majority, I also received a cheque for three hundred pounds along with the news that it was all that remained of my inheritance. My solicitor had invested unwisely and lost most of my money; though I had protested his actions, as a minor I had no power to persuade the courts to force him to invest more prudently. I simply put my head down and pushed forward towards the thing I loved and trusted most: a greater understanding of the natural world. My work didn't have to have meaning or significance beyond that which it brought to my life. It was only much later that I realized it did.

———

The Gaelic word *saoirse* means freedom—and it's freedom of a particular type. *Saoirse* is the freedom to be and to express yourself, the freedom to think and believe what you like; it's freedom of the spirit and the imagination. *Saoirse* and *aimsir*—time—are, I believe, the two most valuable things a person can possess.

The shadow cast by Sunday's Well has stretched across my life and affected me in ways that I'm aware of and, no doubt, in ways I'm not. But I do know that the long shadow that hung over me through my adolescence has made me fiercely protective of my *saoirse*, my freedom to think. I distrust institutions and am a paying member of only one—the Irish Garden Plant Society. I am terrified of being wealthy because I am well aware of how greed distorts people. I have been on the receiving end of this all my life. I am wary of anyone who offers me a good chunk of money for the same reason. Both money and institutions are used to strip people of their freedom, and if they take your mind, they've taken everything.

During my time at University College Cork, a man named Cornelius Lucey was the Bishop of Cork and Ross. (An ancestor of mine, Lord John Beresford, held the same position for the Protestant Church of Ireland from 1805 to 1807. Lucey lasted a fair bit longer, only resigning in 1980, two years before his death.) It's difficult to overstate the power of the Catholic Church of Ireland at that time, and a leader like Lucey—having dedicated his life to a religion that claimed to hold humility, charity and self-sacrifice above all other qualities—was an immensely powerful and a stinking rich individual. He lived in a palace, an actual palace, and his word held weight with the government, the universities, with everyone. In short, he was a man people either looked up to or feared.

One Friday, my tech, Michael, and I were working in the greenhouses when word came that Bishop Lucey would tour the campus the next day. I had planned to work on my own through the weekend to finish taking the week's measurements. "He's not coming into my greenhouses," I told Michael. "I'll be running experiments."

I'd spoken almost automatically, from the gut and not any part of the body typically tasked with high strategy, but I still expected my objection to pass from Michael up through the ranks of the administration to whoever was responsible for making sure the bishop's visit went smoothly. With experiments in mid-flight all across the greenhouse, it seemed to me a nonsensical place to bring anyone not directly involved. Surely the logic of that would be clear to all.

On Saturday afternoon I was bent over a row of pea plants when I heard approaching footfalls on the concrete steps that led from the main biology building. Looking towards the doorway, I saw the flash of a coloured stocking and the hem of a purple robe, then the full robe and the face of the Bishop of Cork and Ross. He must've had a retinue of some kind with him, but I registered no one else. I leapt up and barrelled towards him, yelling from a distance of about twenty feet, "You have to get out!" and pointing him back up the stairs.

But he carried on walking towards me, extending a hand as he drew closer. There was a huge jewelled ring on his finger, and I realized he was offering it to me, either oblivious or indifferent to my directive. He expected me to abandon my experiment to get down on my knees. "I'm not kissing your ring," I said. "And you're getting out of here. There's the door. Go back up the way you came."

He started to get flustered, but I cut him off before he could contradict me and repeated myself, adding that I was in the middle of my work. I was blazingly angry. How could a human being so carelessly disrupt a whole set of experiments? You don't go into a symphony and interrupt the pianist or conductor.

Fully registering my anger at last, the bishop turned and left the greenhouse without a word. I slammed the door after him and went running back to my row of peas still ready to spit flames. By Monday morning, word had got round the university that I'd fired Lucey out, and it was only then that the potential consequences of my actions hit me. My colleagues all predicted that I'd get into terrible trouble, perhaps even losing my place in the botany department. I expected at least a stern talking-to, but nothing was ever said to me by the administration.

A man with even more of the imperious air than the priests at Sunday's Well gave off when they flipped their hems had tried to invade the one place I felt truly free from the pressures and pains of life. And though he was a powerful man, I had defended myself, my work and my *saoirse*. I'd stood up to him and turfed him out, and no one had stopped me or even told me after the fact that I was wrong. In all that, I decided there was a lesson worth learning: guard your freedom to think at all costs.

———

There comes a time in everyone's career when vital decisions have to be made. When I graduated, Ireland did not do much in the way of research. The one place that did, Moore Park, wouldn't take women. So the majority of graduates left for abroad. A lot went to the Arabic countries, England and North America. In the summer of 1965 I, too, had to make my decision, and I did it with a great feeling of misery. In this instance I would have to leave two schools behind: Lisheens and University College Cork, built around my little sumac tree. The medical library exposed the secrets of this tree to me. The sumac was a medicine tree of North America and a spice for the kitchens of the Arabic world. It was hard to say goodbye. But Uncle Pat always told me, "Diana, the world runs in a cycle and you, too, will have your time."

Taken together, my top undergrad marks and my Master of Science from University College Cork elicited opportunities around the globe. I was offered PhD and teaching positions. The University of Florida asked me to be a professor and several South African schools came calling, but I couldn't stomach the racial politics of either place. Ultimately, I accepted an American Federal Fellowship to the University of Connecticut at Storrs Campus.

The fellowship funded a year of individual, private research for a visiting scientist. I studied nuclear medicine—Storrs was the only place I knew of where that was possible at the time—and again kept my subject as broad as possible, examining the effects of nuclear radiation on biological systems. I looked at maximum and minimum exposures in plants and animals; I wanted to understand how radiation could impact the world. In the course of that work, I discovered genetic smearing, in which I isolated a section of a strand of DNA and gently crushed it so its order could be seen and studied under powerful magnification.

I also took some advanced organic chemistry. It was chiefly lab work, and it taught me some valuable techniques for examining and separating compounds.

When the fellowship ended, I returned to Ireland briefly before settling on Carleton University in Ottawa as the place I would pursue my PhD research. There was a professor at Carleton interested in hormones in plants. My master's work meshed nicely with his field of study, so my curiosity pulled me there. Mary Cronin's prediction and the duty to complete my sacred task also added to Canada's draw. For all the joy and knowledge I'd gotten out of my experience at Storrs, the United States frightened me a bit. I couldn't escape a sense—mild but pervading—that everything around me had been militarized. When I encountered police, particularly on a trip I took with some colleagues to Manhattan, I felt like I had stumbled into the presence of soldiers from an invading army. Canada, I thought, would feel more like Ireland, only with the forest left relatively intact. I expected familiarity and natural wonder. I only got one of the two, but it was by far the more important.

The Indigenous peoples of North America are owed a huge debt. Theirs is a magical continent, *an talamh an óige*—the land of youth. The vastness of the place is extraordinary, its wilderness on a scale nearly incomprehensible by European standards, and the Indigenous peoples essentially kept it intact. I feel a huge personal debt to them for that stewardship, for the fact that so much beauty was kept safe long enough for me to bear witness to it. From the moment I first set foot in the woods of Ontario in the summer before beginning my work at Carleton in 1969, I saw Canada, in my heart, as a pristine place, a place of great beauty and bountiful water. It was immediately evident to me that the botanical system here is phenomenal, unlike anything I'd previously experienced. I felt an instant desire to showcase it. I wanted to tell the world that this is a most marvellous place. To this day, neither that urge nor the belief that drives it have left me for a moment.

If anything, my deepening knowledge of the inner workings of the forest only heightened the beauty I saw there. In my PhD research, I focused on serotonin and the tryptophan-tryptamine pathways. Incorporating my existing understanding of medical biochemistry, I expanded my field of view beyond botany and compared the function of hormones in plants and human beings. In humans, the tryptophan-tryptamine pathways generate all the neurons in the mind. The action of some of these biochemicals had been shown in humans, but their existence in trees was not known. Over the three and a half years it took to complete my thesis work, I proved that such pathways existed in plants—in some more than others, and in trees most of all.

Plants contain the sucrose version of serotonin as a working molecule. It is a water-soluble compound in, say, a tree. Serotonin is a neuro-generator. By proving that the tryptophan-tryptamine pathways existed in trees, I proved that trees possess all the same chemicals we have in our brains. Trees have the neural ability to listen and think; they have all the component parts necessary to have a mind or consciousness. That's what I proved: that forests can think and perhaps even dream. This knowledge was new to science. Such connections were not recognized or known at the time.

By the time I finished my thesis and earned my doctorate, I was running out of money to live on, so I looked for work. I found some at the Canadian Experimental Farm in Ottawa with Geoffrey Haggis, head of the Department of Agriculture's Electron Microscopy Centre. There, on a six-month research grant from the Canadian government, I modified the Siemens electron scanning microscope and discovered bioluminescence, a phenomenon in quantum physics. When high-powered energy is shot into an aromatic molecule in a plant, or in the human body, it bioluminesces light. Because the flow of electrons into a given molecule comes out the other side in an equal form of energy (the basis of the famous equation $E = mc^2$), this light can be used as a tracer in any medical field. It can spot, to give one example, cancer in cellular tissue if the light is changed when it is emitted. When Geoffrey and I published the first chunk of our findings, I won an award for the work some time later. (In 2008, three scientists won the Nobel Prize in chemistry for work in this same area.)

We had enough material for two more papers on bioluminescence, but my research grant was coming to an end. Geoffrey wanted to bring me on full-time in order to get those papers ready for publication, but before my hiring could go ahead it had to be put before the board of the Experimental Farm for financial approval. I was summoned to appear before three men. They were all seated behind a single long table in a stuffy room. I felt like I was back in the Cork Courthouse, my fate in the hands of men who cared only for themselves.

My "case," as they called it, was reviewed. Their judgment came fast and final, disguised as advice. Without raising his eyes from the paper in front of him, a board member told me, "You should get married and have children." The board refused to fund my research. Because I was a woman they would not consider giving me employment. They made this very clear to me.

I was back in the lab, mad enough to burn the building down, when Geoffrey walked into the room, fresh from his own meeting with the same board. Without breaking stride, he headed for the glass-fronted box on the wall that held the firehoses and the fire axe. He slammed its door open and selected the axe, calm as anything, and then swung it full force at the box, shattering the glass. In two more swings, he'd hacked it off the wall entirely. With the box well and truly destroyed, he set to work on the lab's parquet floor. He tore up about half the boards before tiring himself out. Sweating, his chest heaving, he coughed out the first thing he'd said since entering the room: "You're not being employed because you're a woman."

"I know," I answered.

"The bastards," Geoffrey said. "If they won't pay for you, they'll pay for this floor."

——

I moved on from the Experimental Farm, sorry to leave the work but happy to be rid of the place. It was 1973 and I was immediately hired by the University of Ottawa as a researcher in the Faculty of Medicine. Over the next nine years, I worked alongside George Biro, a doctor and a professor in the Department of Physiology (now called the Department of Cellular and Molecular Medicine), studying the human heart and circulatory system. Together, we developed a famous stroma-free hemoglobin. This was a non-typing blood cleaned of its stroma, the outer walls of red blood cells that can cause organ damage when artificial blood is transfused into a patient. We also wrote a number of notable papers on cardiac ischemia, which were published by the top medical journals in the world, including the *American Heart Journal*. Ischemia is a condition of oxygen shortage in the heart that can affect any person and can lead to cardiac arrest. On the side, I also completed a one-year course in general experimental surgery, giving me a greater understanding of the body and its circulation.

The work was fulfilling and deeply meaningful to me, and I knew it was having a positive effect on people's lives. Much later, a severely ill friend came to visit, bringing her medicines with her. I saw her, in my own home, give herself a treatment in which the life-saving drugs were delivered to her through the medium of one of the artificial blood products with which I had worked. That moment is engraved in my mind still: seeing someone who benefitted directly from my research left a lasting impression on me.

Despite everything that was going right in my life, though, I faced huge frustrations—frustrations that would be familiar to any woman making a go of it in professional life in the 1970s and '80s and that, sadly, are still familiar to most women. Science and technology faculties and workplaces are still known today as particularly rich sources of insecure and misogynistic men. My experiences were both unique to me and true to stereotype. I even had ideas and discoveries stolen and passed off as the work of my male colleagues. I had so many other painful experiences that, finally, I'd had enough.

Luckily, by that time, I was truly, happily married. I'd first seen the man who would become my husband, Christian Kroeger, at a gathering I'd organized while still at Carleton to introduce a then unknown scientist, David Suzuki, to various university-associated big shots and socialites. The crowd was formal in dress and personality, and the whole thing was pretty dull until, partway through the evening, a pair of men crashed into the room. Fresh from a caving expedition to study a rare species of bat, the new additions were still wearing their coveralls and boots, with a barely dried layer of mud on top. I got a kick out of them. It turned out I especially got a kick out of Christian, who I later learned was the son of a NASA engineer, one of the key architects of the Apollo space program. Christian, having himself worked in the United States space program, moved to Canada in his twenties, where he got a job in the federal civil service. He wrote policy and legislation around projects as important as the Access to Information Act and the building of the Confederation Bridge that links Prince Edward Island and New Brunswick. An active caver and climber, he'd gone on the bat expedition as a favour to a friend.

Christian and I got together, bought a piece of property, got married and got to work, each with our hammer, on the farm where we've now lived for more than forty years. In short, at home, there were plenty of plants and one very fine man.

So when I reached the point where I just couldn't take it anymore, I came home to Christian and told him that I was sick of the bullying, sexual harassment, backstabbing and pettiness endemic in academic science. We'd already built a house and put in gardens and an orchard on land that had been an empty field when we arrived. When he asked me what I was going to do instead, I said, "I'll just do my own research."

My Own Work
in My Own Way

WHEN I WAS still very young and living with my mother at Belgrave Place, I snuck away from the house one day and went to Woolworth's department store. I'd been given thrupence, I can't recall why, and the trip was my first ever adventure out into the city alone. Moving through the islands of merchandise, I landed at last at a display of seed packets and selected a small waxed envelope of Black Seeded Simpson lettuce seeds. I paid for them with my thrupence, the man behind the counter giving me a strange look because I was so young and obviously out of bounds. I tucked the packet into my safest pocket, then went back home and hid my purchase.

On a raised terrace behind Uncle Pat's house, there was a common area that held the neighbourhood's vegetable plots. The next day, I headed for it with my packet of seeds. The commonage was surrounded by a stone wall topped with broken glass and the only gate was locked. So I threw my cardigan onto the top of the wall to cover the glass and scrabbled over. I half expected to drop into a jungle, a Garden of Eden with trees heavy with apples. Instead I found individual plots bedded down together for the colder months. Still, I was not dissuaded. I found an empty patch of soil and made a crooked furrow in the earth with a stick. I poured my whole packet of seeds into that row and covered them over. Then I set to work to create a viewing post from a nearby stump. Around it I laid out pieces of a bush that had broken off in a windstorm so that I had a camouflaged place from which to monitor the progress of my lettuce without being seen myself. I returned day after day for a year or more, talking to my seeds from inside my false bush, asking them to grow, asking what I could do to help them.

Of course, the seeds, planted at the wrong time, didn't sprout and the ground never seemed to change. But that effort was the first manifestation of my desire to have the living world respond to me as a person, to see my efforts recognized in the form of green growth. That desire has always been with me, like a second heartbeat. And in Canada, with Christian as my partner in the endeavour, I could finally seek out the land that would allow me to drink in the living world to my heart's content.

We'd been looking for a year already when we were shown the piece of land in Ontario that would become our home. It was the early 1970s and between us we had six thousand dollars and a shared refusal to take on debt. The listing, for sixty acres, was in our price range. (We would later add a neighbouring hundred-acre tract to stave off developers.) When we drove out to the address, we found a huge field, lying fallow, with a slight south-facing slope to the land. I got out of the car and immediately felt a sense of peace, like I was being drawn into a welcome embrace. In all the other places we'd seen, and even the ones we'd seriously considered, I'd never felt anything like it. It felt like the ground was calling out to me to say that it would be good to me. I listened and we bought the place. Every day since the embrace of this land, and its bounty of surprises, has held me close, nurturing me to be something that I don't quite understand. I have nurtured the land in return, and that energy has only amplified in the form of peace.

After we took ownership, Christian and I walked the land to get a proper sense of things. We ventured into the cedar woods, where I had my first direct experience of a truly massive tree. It was dead, the colossal stump of an eastern white cedar, *Thuja occidentalis*, that had been wrenched from the ground. The rest of the tree, which must have been the size of the colossal redwoods still living on North America's west coast, had been cut down and dragged out. But the stump was too big and unwieldy, its bole so large it couldn't even be burned.

Seeing this last vestige of the Canadian virgin forest that had once stood on and all around our land, I felt the same mix of awe and grief I'd known when Pat Lisheens showed me the heartwood of that ancient bog oak while cutting turf. *My God, this is what was here before we came,* I thought. And then, *What were they thinking of to cut those giant beauties down?* The stump before me wasn't living, not in the way the tree it belonged to had been, but it *was* alive. Those big boles still take in water; they change shape; there is a form of life still present in them. The majesty of the forest that once stood on this karst land outside of Ottawa is enough to stagger me. As is the arrogance and greed it took to cut it down.

We set to work building a house for ourselves. To save money, we bought planer outs of white pine tongue-and-grooved from the local lumber mill—one reason there's no other house like ours on this Earth. During the construction, we lived in a shed in the front garden roughly the size of an ice-fishing shack, and not all that much more luxuriously appointed. Weekdays, I'd head into the medical labs of the University of Ottawa early, shower there and get myself all fluffed up for the day. After work, I'd drive home to the shed. In the winter, we had a small kerosene heater to keep us from freezing at night. As it got properly cold outside, the deer mice would come in and join us. They'd settle in a toasty spot under the heater, and we'd get peanuts to feed them. We didn't have electricity or indoor plumbing, let alone a television set, so that was our entertainment: watching the deer mice nibble peanuts in the glow of a kerosene heater.

Just a few years earlier, when I worked at the Experimental Farm, I used to take a walk on my lunch break that brought me past an area where there was a tulip tree growing, a beautiful *Liriodendron tulipifera*. It was the only tulip tree on the Experimental Farm and I loved it. During one summer thunderstorm, it was hit by lightning and damaged too heavily to save. On my walk the next day, I saw what had happened and talked to the groundskeeper who was cleaning up the aftermath. When I told him how sorry I was to see the tree lost, he answered that I didn't know the half of it. "That's the last tulip tree up here in Ottawa," he said.

I was dumbfounded. I hit the stacks and discovered that *Liriodendron tulipifera* was a medicine tree to the Huron people, who considered it to be magical and sacred. There is a quinone structure in the tree, giving it anti-microbial and anti-parasitic properties, and the Hurons used its wood it to make their death masks. I also discovered that not only was the groundskeeper's claim true, but the problem was also bigger than just that one species—many trees were on the brink of disappearing around us. I had always kept a list of species I hoped one day to grow, my gardener's bucket list. After the loss of that tulip tree, I began to add threatened species as well. I also went down to southern Ontario, got two tulip trees from a nursery that sold native species and planted them.

Christian and I married on September 14, 1974. Given what had already happened to me in life, I wasn't comfortable pledging to marry "for better or worse," so I made the clergyman conducting the ceremony pledge us "till love do last." For a wedding present, my colleagues at the University of Ottawa purchased me an orchard's worth of trees. One researcher spoke for the group. "Diana, we don't know anything about the plants that you're interested in," he said. "So you give us a list for your wedding." The best tree they found for me was an apricot tree from Manchuria, *Prunus armeniaca*, a hybrid called the Harcot apricot. It is still going strong.

My master's work had highlighted to me the importance of focusing on hardy species, particularly given Canada's climate. The trees on my orchard list came from all over the world, but I wanted northern genes, the genes of survival. (Also, I discovered, these northern trees offer a higher ratio of natural medicines.) I knew I could then crossbreed these strains into plants whose resistance to drought and swings in temperature I wanted to increase. The group gave me pears, apples, peaches and plums, and different kinds of cherries, which arrived as seedlings or three-inch slips. I planted my Manchurian apricot tree next to the three-quarters-finished house knowing the radiant heat given off by the building would be enough to keep it alive, something called the black-box effect. In addition the radiant heat allowed pollination to occur, a small experiment that meant the tree fruited and I got to eat apricots. In all things, the idea was to expand the truncated Ontario growing season wherever possible—through frost resistance and any other means I could dream up.

The vegetable garden had gone in as soon as we bought the land, a financial necessity as much as anything but a distinctly pleasurable one. My sole membership to the Irish Garden Plant Society allowed me access to a wide array of heritage seeds. I found native hazelnuts growing wild on the property and worked with them. Towards the end of the decade, with the house now built, Christian and I also brought in many species of black walnuts, some from neighbours' properties, some from Gananoque—on Lake Ontario, about an hour southwest of us—and some purchased or given to us, even left to my safekeeping in wills.

If we both had a spare stretch of time, we would set out on plant-finding expeditions in the area around our place in eastern Ontario. I set up criteria for identifying top-quality trees, many of which, I'd noticed, were falling under the chainsaw at the time. I keyed in on qualities related to the tree's size and health, as well as its provenance and the logistics involved in preserving it—when it would seed and whose land it was growing on. I wanted trees that were as close as possible to virgin forest because their genome is the healthiest, though unfortunately only the runts have been left behind in many forests. These trees, from forests that in Canada have a track record of thousands of years, are the best of the best I could find: the species most perfectly suited to the climate and best equipped to fight disease.

To identify potentially rich areas for our searches, Christian and I studied topographical and aerial maps. All of the cutting that was done when people first settled happened on the "good" land—flat, arable tracts that were good for farming or easy to build on. Any stretch with hills or big rocks or surrounded by swamp was too much trouble for people to get at, so it was more likely to have been left alone. These were the haunts of quality native trees. We also scouted century farms, properties that had been settled 100 to 150 years ago, as the name implies, and ideally kept in the same family. When I spotted a quality tree—sometimes yelling for Chris to stop the car so I could leap out for a closer look at something growing on the side of the road—we'd find out who owned the land. If they had something really special, we'd meet with them and tell them about the tree and its importance.

I remember a woman called Mrs. Fern, an elderly lady who owned a piece of land on the St. Lawrence River. She had the biggest shagbark hickory we had ever come across. I knocked on her door and she answered. I told Mrs. Fern, who lived alone, that in all our travelling, which was considerable even then, hers was the most impressive shagbark we'd seen and that I thought it could be from virgin forest. The tree grew between the road and the powerline, and Mrs. Fern told us that just weeks before our visit the hydroelectric company had wanted to take it down to allow easier access for their bucket trucks. She had asked for time to think about it. "I had a feeling I shouldn't do it," she said, "and now I'll tell them not to cut it down."

Mrs. Fern invited us in for tea. She was absolutely thrilled with the news of the tree and with my request that she safeguard it, and even include a clause if she sold the property demanding the buyers do the same. Finding out you've got a special tree can really buoy a person up. That shagbark hickory is still standing today, a lone survivor of the bucksaw.

Oftentimes, I asked permission to take seeds or cuttings from a tree that caught my fancy. I was never refused. I took those samples back to the farm, segregated them to keep them quarantined until I knew they didn't carry diseases, and set up growth trials with them. The species that performed best, I then planted permanently on the property, developing a collection of the best of the best rare and native species. This was my arboretum of Canadian trees.

A century ago, the area we live in was quite poor. Many farm families grew really big orchards, and peeled and dried or otherwise preserved the fruit to have something to eat through the winter. The apples were Russian species, northern bred, adapted to the cold and crossed here. They also had some pear and plum trees, though not many, and sour cherries, which I know for a fact came over from Ireland. Some places had interesting crabapple collections to boot. One common crabapple, called the "water core" by the local farming community, was a delicious small apple. I found one tiny tree, which I wasn't able to propagate, and that tree died. (I fear the variety is now lost.) When Christian and I were looking for fruit trees to add to our orchard, we searched for house bottoms, the remains of old foundations, in deserted areas that were marked on the older maps as farmland. And I got all of my roses from cuttings I took in forgotten graveyards. People in the past tended to plant rosebushes near the headstones of their loved ones.

I also started what became my North American Medicine Walk, collecting lost medicinal plants. The biochemical philosophy around this area of the garden is based on aerosols or volatile organic compounds (VOCs) released from plants—the scientific basis of many ancient Indigenous medicines. These healing properties are of particular interest to me as a scientist. Top of the list of plants I wanted to find was a lost peony that was black and smelled of chocolate. There was one left in the world, in Oxford University's collection. The peony, *Paeonia officinalis*, is an important plant of traditional Chinese medicine, a potent vasoconstrictor also used for ridding the bladder and kidney of stones. After ten years of looking, I sent a pleading letter to the university, and soon a rack filled with test tubes, in which agar-agar had been used as a support medium, arrived at the post office, and ten little somatic clones lay one on top of each test tube, scraped from the mother plant in England. I managed to grow these green babies on. They flowered in four years, and the flowers were all different, doubles and singles; some shades were nearer to black than others. I crossed these darker plants to produce a colour even more black and they are the heart of my peony collection. I visit them when they bloom every June with a cup of tea in my hand to cheer them on.

My searches and the expeditions I took with Chris aren't the only way plants make it onto our farm. The towering oak I now admire out my kitchen window was planted by squirrels twenty-five years ago. I'm also willed plants. The first time that happened, I was left a gooseberry bush that had been developed at the Experimental Farm. The man who passed it down to me was a breeder there. Until recently, I wrote for botanical gardens across the globe. I took payment in kind, not cash, searching through their listings for plants to repatriate, then asking for seeds or cuttings. I always looked for surprises and I still do. In Canada, I also consulted Indigenous people; descriptions of what they would look for in a given area directed me to many plants I would never have found on my own. (I am proud that some of them have called me the "medicine keeper.")

Older farmers were another rich source of knowledge. A neighbour named Grant Baker told me that one particularly lean winter in the 1930s, he loosed his entire flock of Cheviot sheep into a forest of cedar— the eastern white cedar, *Thuja occidentalis*. He didn't have enough money to afford their feed and didn't want them to starve to death in his barn, so he ran them into the trees. The sheep, he said, survived on the cedars, and came through in the spring looking as healthy as ever. I looked into it and found that winter sparks a chemical change in the cedars that produces a vanillin-like compound called delta-fenchone, a natural taste enhancer. It's usually enjoyed by the deer, who do seem to find it very tasty, but all hoofed mammals can eat the green leaves of cedars for the whole winter season. Later, I discovered that the cedar was called the Tree of Life by Indigenous people for that reason: it has cones for the songbirds and green leaves for the ungulates and rabbits.

Friends have also gone to great lengths to get plants to me. I have a white cherry that I thought was a lost varietal until some German friends, Heidi and Norbert Weiler, found one on a farm in the upper Alps. The elderly couple who owned the farm planned to have the trees cut down until Heidi and Norbert asked them not to. "I know a woman in Canada who's looking for these kinds of cherries," Heidi said. I sent along instructions for preparing the fruit for shipping. These white sour cherries have to be fermented on the fruit-stone before they will germinate and grow into cherry trees. I now have the species ready for reintroduction.

The element of luck, or fate, in the rediscovery of that white cherry, and in finding Mrs. Fern's shagbark hickory—getting there just before the plant is cut and potentially lost forever—is common to the story of many of the species that make their way to our farm. It's almost as if I put a searching energy out into the world when I'm hoping for a species, and by some serendipity the species answers and comes through my door just days or weeks after I make my wish.

That happy ending didn't seem possible in the case of the hop tree, *Ptelea trifoliata*. In appearance just an undistinguished little tree, not much more than a shrub, the *Ptelea* once grew in the forests of Ontario and was a sacred tree of the First Nations. When the pioneers first set-tled, cold and battered families desperate for firewood, they were told by the colonial authorities that to own their land they had to cut down all the forest on it. They usually kept a couple of big maples around because they'd heard from the Indigenous peoples that you could draw syrup from them. The fire-keeper of Akwesasne Mohawk Nation told me that the settlers were also offered a piece of advice about *Ptelea*: Do not cut down this sacred tree. To this wisdom, they didn't listen.

All the First Nations that knew of it used *Ptelea trifoliata* as a traditional medicine. Despite its sometimes shabby appearance, that tree has a synergistic biochemical that revs up your major organs and causes them to metabolize things faster. It allows your body to make efficient use of medicines, magnifying their potency, and reduces the amount that you need to take. For instance, in modern medicine, if you need chemotherapy and took the drugs with *Ptelea* as a companion, the dosages would be reduced substantially. Such medicines are known as synergists. When there are less chemo drugs in your body, for instance, you are better able to cope with the side effects, to handle their stresses and excrete them and heal. I'd had a variety of this little tree on my bucket list for about twenty-five years, but it seemed to have been completely extirpated in Canada and possibly driven to extinction. I had almost given up trying to find it.

Around the year 2000, I was doing a radio show for the Canadian Broadcasting Corporation that led to a series of programs slated to run on radio stations all over Florida. In my research, I'd discovered that the Seminoles used *Ptelea*, too, and so I figured that Florida might be the last place it still existed. Hoping that an energetic listener might go out and find a *Ptelea trifoliata cressidifolia* in the wild, I gave instructions over the radio. I described what to look for and what I thought a mature tree might look like. There were no existing visual descriptions, but I projected as best I could. I also described when to look: from the end of July on into the fall, the trees carry a distinctive seed pod, round and thin like a large coin or a communion wafer—hence the tertiary common name, wafer ash. I then sent the description to naturalists all over the eastern United States. I got responses to both the letters and the programs to let me know that loads of people were looking, but none of them succeeded in locating that variety of *Ptelea*.

After those efforts came up empty, I assumed the tree was lost. That was a terrible feeling. You can't remake a living thing. When it's driven out of existence, it'll never come back. That tree had a huge value in the past and I thought it would have a huge value in the future. But instead of that positive power to help and to heal, we had a subtraction, a hole in the shape of the tree and the medicine that went with it.

About five years later, I was in Fort Worth, Texas, where I'd been asked to give a week's worth of talks about climate change to the local schoolchildren. While I was there, I was sent an invitation from a very wealthy woman whom I'll call Maria, asking me to join her for morning tea. I agreed, but the morning of the appointment I nearly cancelled. I was staying with my good friend Ginger, who'd put me up in a four-poster bed so big it had a little set of steps to climb up into it. When I woke up I forgot about the steps and fell down them, ripping up my legs and causing me to swear like a Trojan. I was meant to arrive for tea at eight o'clock and didn't want to go with my torn-up legs. My friend convinced me to tough it out.

A chauffeured car arrived and took me to the tall iron gates of a huge estate. The driver hit a button and passed through the gate when it opened. Inside, there were men everywhere decked out in tactical clothing and carrying guns—Maria owned a private army.

A butler let me in the front door and directed me to the library. The room was predictably large; Van Gogh, Dali and Renoir hung on the walls. Maria was there with the tea waiting. She offered me a seat and a large, antique famille rose plate with a biscuit on it. I thought about the fact that I couldn't afford to break the plate, a rare Chinese one, and set it on a table, eating the biscuit off my palm.

We made small talk and Maria remarked on my name. "I used to have some neighbours who were Beresfords," she said.

"Where was this, then?" I asked.

"In Arizona and New Mexico."

"Those were my relatives." A thought hit me. "So you own land in New Mexico?"

"Oh yes."

"And how many ranches do you own?"

"I have six."

"You do, do you? Do you happen to own a mountain?"

"I do."

"Is the mountain gravelly?" My wheels were really turning now: *Ptelea* has to grow in gravel. The roots of the tree need more oxygen than many other trees.

Maria confirmed that her mountain was gravelly.

"Have you ever cut the forests on any of the ranches?"

"No, it's all as it was."

Maria's family had owned the land for many years. Correctly guessing that I would be particularly interested, she told me that at one point, many years ago, they'd had a botanist down to the various ranches to classify as much plant life as possible. The botanist had arranged a herbarium that she still kept in its own house on one of the properties.

I was so excited I could barely get the question out: "Do you have *Ptelea trifoliata cressidifolia* in the collection?"

She didn't know offhand, but told me she'd have one of her property managers check. "We'll know by tomorrow."

I got a call the next day. "I'm inviting you up to the house," Maria told me, adding somewhat anti-climactically, "I have champagne ready."

I was ferried back to the house and let into the library. Maria was there with her champagne uncorked. Three sons and several grandsons were arranged in a line, the smallest boys with dicky bows. Everyone held a glass. Each of the boys approached and shook my hand. Then Maria said, "I have to announce that we have *Ptelea trifoliata cressidifolia*."

"You're joking!" I screamed. I would have jumped around the room, but there were too many children and expensive objects to risk it.

Maria gently brought me back down to Earth. She told me the herbarium, which held the specimen proving the *Ptelea* grew on the property, had been collected fifty years earlier. Though they'd never cut the native growth on the ranch it had come from, there was still no guarantee that a *Ptelea* would still be growing in the wild. I was set to leave Texas in a couple of days. Maria asked if I'd be willing to return. "Come back and we'll go down to my biggest ranch and see if there are any still there," she offered. Of course I accepted.

It was a matter of months between that glimmer of hope and a point in both our schedules that would allow the exploratory follow-up trip. The waiting period was a roller-coaster. Every day I experienced a period of hope, where I was sure the *Ptelea* was all but found, and a period of utter despair, where I couldn't shake the surety that it had been lost forever. The *Ptelea* even entered my dreams like some mirage that kept disappearing every time I drew close.

I went back to Fort Worth and stayed with Maria for a night. The next morning Allan the pilot flew us to her ranch by private jet. Then I learned that Maria had hired a helicopter. The pilot, a Vietnam War veteran, took me out to search. I rode suspended face-down out the side of the helicopter in a rigging of leather straps, wielding a pair of heavy naval binoculars for a better view of the ground. I was strapped to that helicopter for four full days as we gridded the whole ranch, which was two hundred square miles. On the last day there was a heavy dew in the morning, which doesn't often happen in New Mexico. That dew got caught in the circular web of the *Ptelea*'s seed casings and made them reflective to the sun. Even from our cruising height, it was impossible to miss, as if the tree was holding handfuls of shining coins. "Let me down!" I screamed to the pilot over the radio. I was so excited, I'd forgotten for a moment that I was up in a helicopter.

He found a place to set us down, neatly, on a mound of gravel. I bounded from the helicopter, forgetting all about the risk of snakes. I ran straight to the tree and wrapped my arms around it and started bawling. I just cried and cried.

Back at the ranch house, I told Maria I'd succeeded and she came back out with me to see it. It was late afternoon by that point and the dew had evaporated. Without it, the tree was almost invisible on our approach, indistinguishable from the brush around it. There was only the one fully mature *Ptelea* on the whole ranch. Maria hauled out the celebratory wine again and, after a glass, I told her she could now consider me a literal tree-hugger.

Splitting Wood

I LEFT MY job at the University of Ottawa in 1982. By that point, eight or nine years into our residency, the farm was starting to take shape and our house had begun to feel properly lived in. The gardens were expanding out towards the eight acres they occupy today. The orchard was starting to produce fruit in earnest. We'd planted the black walnut trees in a great ring, an *allée* tracing the roundabout in front of our house. Hedgerow plantings around the gardens and fields increased the buffet of nectar and pollen, encouraging birds and insects to come and stay awhile. And we maintained the whole 160 acres organically, with dormant oil sprays, sulphur sprays, whitewash and the like—a true rarity in a time in which every farmer's field in the area was absolutely dripping pesticides. My whole life I'd had a wild love of nature and a desire to be surrounded by green growing things. Here was a place where I could work with my beloved plant kingdom every day.

My father-in-law, Hermann Kroeger, was Deputy Director at the Marshall Space Flight Center and an integral part of the Apollo Program. He told me a story from his university days that has stayed with me. He had started a practicum for his master's degree in aeronautical engineering, learning about some of the materials with which he might one day be asked to build a rocket or a plane or something else people trusted with their lives. His professor presented him with a barrel of metal castings and a file, and a simple assignment: grind all the castings down by hand. They had to be perfectly smooth.

It seemed at first like a mindless, even cruel, bit of homework. There were times over the days he spent filing the castings that he believed the whole endeavour to be little more than a professor's power trip. But he emerged on the other side with sore arms, a pile of metal filings and polished castings and, he assured me, a more complete understanding of that material than he could ever have had without having completed the assignment. Total immersion gave that to him, along with an appreciation of the material he could have gained no other way. This simple task played a key part in getting a man to the moon. On the farm, I sought my own version of the metal filing effect. I wanted to immerse myself in nature to the point that I understood it as intimately as the people I love, as the inside of my own head.

Surrounded by nature, drinking it in like air, there was no telling where a discovery might pop up. As I watched Chris bringing his axe down to split a dead tree for fuel for the wood stove, I might see the same dark swirls playing across the bark of piece after piece. As I stacked the split wood, I'd write this off as rot, an assumption that was correct much of the time. But eventually, having examined untold thousands of pieces for the woodpile over many years, seeing such patterns again and again would start to bring out small differences. A streak of purple, pink or yellow would catch my eye and suddenly I'd realize that I wasn't looking at rot; it was a fungus growing not just on a tree but also in it. I also noticed that each particular species of tree had its own group of two or three specific fungi that was different from the fungi found in other trees.

What did this mean? The fungi were all from the higher order of their kingdom, more evolved than other forms of fungal life and very sophisticated in the manufacture of complex biochemicals—many of which are too difficult to copy in a laboratory. The fungi use these compounds in an assault on the tree, and the tree reacts in self-defence by producing its own array of chemical actors. All of these substances, which I studied under my microscope, had medicinal properties. So the product of this warfare was medicine and the trees' release of beneficial aerosols into the atmosphere. Now clinical studies are showing than many of these tree species boost the immune system and shield the body from a variety of cancers.

You have to cut down the tree and go to the woodpile and split it, over and over. You have to see the world with your own two eyes to notice that there's something going on, something you don't really understand. Wanting that immersion was part of the deep desire I'd felt to own land in the first place. And the observations it yielded, the seemingly small things I noticed and the desire to understand why they were the way they were, formed the foundations of the most significant ideas of my life.

I also brought things with me when I headed out to the gardens, the fields and the forests. I had the whole of my scientific education, a baseline of knowledge and technique against which I could measure and make sense of the things I observed. I had the freedom of being loosed from institutions, the room to be myself and chase what caught my curiosity without external interference. I had Christian's love, belief and support, and soon that of our daughter, Erika, which were sources of unbelievable strength. I had the solitude and desperation of existence on the margins, scary and sometimes painful emotions that have the power to strip you down until you're just your raw essence and nothing more. I had my painter's eye for beauty, which could draw me to a discovery just as surely as the orderly thinking achieved by reason. And I had the ancient knowledge of Lisheens and the view it had given me of nature as the sacred source of everything that's needed to sustain ourselves and the planet.

What I didn't have was much in the way of resources. Anything I wanted to investigate had to be cheap, so, as I had done when I selected species for my master's work that would fit into the growth cabinets, there were logistical parameters around what I was able to pursue. But those exist to one degree or another for any scientist, and ingenuity could always stretch a dollar and an idea. I was also fortunate that the world that was observable for free on the farm never stopped providing me with questions I was burning to answer; the property itself was the ideal laboratory for putting my ideas on trial.

Early in our residence, Christian and I gridded the farm the way I later would Maria's ranch and searched it for old boles, the last vestiges of virgin forests in the area. We found the great stump of the *Thuja occidentalis* in that process, and outside of the cedars, I noticed that many of the former giants we encountered were hickories. Their size was just stupendous, boles ten feet across, trees so big that the men who clear-cut the area all those years ago couldn't really handle them and had struggled to transport them out. That they had once grown so well was a fair hint that they might be well suited to the area. With my curiosity sparked, I started to study hickories. I looked at the tree's weight, wood volume and ability to sequester carbon dioxide, and discovered that in this part of eastern Canada, hickories stood alongside oaks as the top species for carbon sequestration.

I cannot say enough to praise hickories. They are an extraordinary species. There are about twenty of them, mostly in North America; they even go down into Central America, with one poor orphan in China. They are an exact match to the long face of this continent, where they can suck up the excesses of solar radiation. Hickories were traditionally the lifeblood of Indigenous peoples, who made a kind of cheese, an oil, a milk, a cream and an alcoholic beverage all from their nuts, and consequently did not appear to suffer the multitude of brain diseases we have today. Hickories' first-class status in the fight against climate change is a result of their great need for atmospheric carbon dioxide to thrive and produce huge crops of high-value nuts. I have planted almost all of the *Carya* species in my arboretum but I am still looking for *Carya tomentosa*, a hickory used by the Indigenous people of the northeastern United States and southeastern Canada to smoke meat and vegetables in order to preserve them. The kingnut, *Carya laciniosa,* is my all-time favourite tree.

In watching the trees I planted over time, I learned things first-hand. I had long known of the inhibitory effects on other plants of certain chemicals produced by walnut trees, for instance, and saw empirical proof in my arboretum. Even planted as far as sixty-five feet away from a walnut, susceptible trees such as apples, pears, plums, and hawthorns will succumb to those chemicals—though it took almost thirty-five years for this to happen. As a result I lost some favourite hybrids; even certain cedars were affected.

Observing wildlife behaviour led to similar discoveries and some longstanding practices. We left large stretches of our fields unplanted so the birds would have a place to enjoy the dust baths that keep mites and lice out of their feathers. Every March, we clean out all the bluebird and tree swallow boxes and fill them with fresh wood ash so that the travellers arriving after long migratory flights can fluff their feathers and enjoy an indoor bath. Surrounding some of our fields and gardens are dry-stone walls, against and within which snakes, newts and salamanders will happily overwinter, safe from predation. And ahead of that coldest season, I rake small piles of leaves against the faces of the walls, like colonies of little tents, for ladybugs to shelter in. And whoever knew that porcupines, cedar trees, and deer have a lifelong connection. During the winter when there is snow on the ground, porcupines will leave their dens during the night to cut branches from a cedar. They'll eat a few and leave the remainder on the ground for hungry deer to eat.

Along with the planting of native and rare species, these smaller, precise efforts also meshed together to serve a larger goal, that of extending a welcoming hand to life in all its forms. My gardens would never be a place of sterile beauty, with the intricately beautiful faces of flowers gathered from around the globe looking out at the rest of the world from behind a fence of privilege. Whether it was with me at birth or cultivated in Lisheens, my love was for an active and open beauty, the wonder that exists in the intricate relationships between all living things.

For instance, I have a willow that was planted from a three-inch slip twenty years ago and is now already a tall and gangly tree. Early this past summer, I noticed a row of holes—tiny, just a sixteenth of an inch across—in the bark of the tree about six feet above the ground. They were clearly the work of *Sphyrapicus varius*, a species of small woodpecker called by the common name of sapsucker. The yellow-bellied sapsucker, the particular species native to eastern Canada, is considered a pest by many gardeners, and with good enough reason. The sapsuckers open their boreholes so that sap, the sugar of the tree, will collect at the edges, allowing them to live up to their names. They inflict these wounds on live trees. Particularly when there aren't enough suitable sources of sap, the damage can result in the death of the tree.

Rather than devise a way to drive off these birds, though, I watched and waited. Two weeks after the sapsuckers had done their work and eaten their fill, butterflies began to appear in great numbers at the holes. The interval had allowed the sap at the surface to crystallize. I realized the butterflies were there to have their own feed of sugar and also to enjoy electrolytes from the tree that they would have had no access to without the sapsuckers' hard work. These electrolytes paint the colours into the butterflies' wings.

Already, I had been treated to a slow and beautiful dance, and when the butterflies departed, I saw that they left behind the music. The sapsuckers' boreholes are the exact size of opening in which ichneumon wasps prefer to build their nests. The ichneumon is a parasitic wasp that keeps a variety of nasty pathogens out of a garden—a hugely beneficial insect. Nature is very forgiving, in its way, but it can only be interfered with so much. Had I not followed the row of holes for the entire summer, I wouldn't have understood their significance or have been treated to the full, enthralling chain. Had I gotten rid of the sapsuckers—as many gardeners would have rushed to do—I would have potentially lost the butterflies and wasps. There is never a day spent outside that you don't learn something. It might be something small, but that small thing might also be a key to something very big. The discovery of those small things and of the ways they connect to one another and ripple through the whole web of life—that is one of the true beauties of nature. That is what I sought to understand and cultivate.

I can say now that the whole of our 160 acres is designed with the goal of encouraging life in just these kinds of ways. The outskirt plantings of our hedgerows bring in birds and insects, which find ample food and lodging when they arrive, as well as a reprieve from the chemical onslaught they face nearly everywhere else. Details like the placement of the boxes on our bluebird walk or the decision to leave the sapsuckers alone mesh into a cohesive unit. Looking at the whole, it would be easy to think that the place had been laid out all at once from a single master plan. There is truth to that. The design did tend toward nature's inherent plan from the beginning of my tenure on this land, but from my perspective on the ground and down in the dirt for all those years, I didn't always see that.

I started saving species simply because I thought they were important and shouldn't be lost. When I placed special emphasis on frost and drought resistance, I did so because my master's work and the lesson of those in-between species I learned about while rewriting the third-year botany lectures had left me in no doubt that forest loss would lead directly to a changed climate. With no sign that clear-cutting of the world's forests could be contained, I had no reason to believe we weren't headed for a future in which only the hardiest species could hope to survive. Biodiversity as we understand it would be a thing of the past.

Many other elements of the farm design were shaped by trial and error, among them the safe distance that now exists between the orchard and our walnuts. Like any gardener, I tried many things that didn't work. I had a species of peach called the Reliance that was capable of surviving through the Canadian winter, but the material I grafted it to was not. When the cold weather killed the root-graft union, I lost the

Reliance peach. I learned a valuable lesson from that loss. I also know that a winter-hardy peach is a very real possibility, though there isn't time left for me to develop it in my lifetime.

We had lived on the farm for the better part of a decade when the vision of the property as a unit at last became fully clear to me. I woke up one morning to the sight of a blood-red male cardinal looking in my window from his perch on a limb of the apricot tree, which was covered in pink, star-like blossoms. As I wrote in the introduction to my book *A Garden for Life*, I found myself in a silken moment: seeing for the first time the totality of what I'd been building in partnership with the natural world outside my bedroom window.

"Bioplanning" was a term I coined to describe it. Like many things in nature, it is a simple concept that opens up into endless complexities. As I wrote, the bioplan is "the blueprint for all connectivity of life in nature." It is the web, both seen and unseen, that ties the willow to the sapsucker to the butterfly to the ichneumon wasp—and ties all of them to us. It is the evolutionary framework, the balance, the give-and-take-and-give, that allows life to exist and thrive on our planet. Bioplanning is the act of aiding and encouraging the bioplan. In a garden or on a farm, that means realigning the garden to encourage its use as a natural habitat.

Seeing for the first time how the whole of the farm meshed into a single bioplan was a sacred experience. The fundamental truth of nature exists around us at all times. By keeping myself open to it and responding when I caught glimpses, I had built, piece by piece, something substantial that is right and true from its tip to its tail. And there were bigger, global lessons, and even a road map, visible just beyond that initial realization, lessons that wove together the experiences of my whole life, back to my childhood days in Lisheens.

———

Climate change is the biggest challenge humanity has ever faced. It touches every living thing. It is a daunting task to even see the entirety of it. To try to think of solutions can, for any individual, quickly begin to feel like an impossibility. Because the problem is so big, many people simply turn away; they deny it exists at all. Others acknowledge the truth being shouted out by every climate scientist on Earth only to embrace the cynical belief that people will never change and there's little point in trying to save the planet—we're all doomed. I'll put it plainly: I have no time for anyone prepared to turn away or throw in the towel.

On the day Pat Lisheens and I headed out to cut and stook the five-acre barley field, that task seemed too big to properly comprehend. Of course, compared to confronting the threat to life we all now face, two people tackling a field of grain is a small thing. But inside your head and heart, when a problem seems beyond your ability, it doesn't much matter how far beyond—impossible is impossible. That day taught me to take the first step anyway. It taught me that our limits, alone and together, lie much further out than we imagine. It taught me there is no such thing as a hopeless situation.

That silken moment of realization that every effort I'd made on behalf of plants, animals and insects had meshed with every other, compounded and multiplied, gave me the same feeling of empowerment as my day in the field with Pat. I was approaching from the other direction—instead of seeing the goal but not how we'd achieve it, I'd seen the steps without comprehending that they followed the same road. But the lesson was the same. Positive action, no matter how small, builds towards your bigger goals. Like the pieces of knowledge given to me in my Brehon wardship, every effort to aid and encourage the natural world is as valuable as every other. Whether we are the mighty or the meek, we must all act to stop climate change. We are all siblings in the communal family and the natural world is our commonage.

In the years since that initial realization, I have developed a course of action for halting climate change. I call it the global bioplan, a patchwork quilt of human effort to rebuild the natural world that will envelope the entire planet. It is not the ultimate solution to climate change; it is a means of reversing the damage done and of buying us time to find that solution, of stabilizing the climate long enough to address our destructive behaviours in earnest.

The core of the global bioplan is a simple idea. If every person on Earth planted one tree per year for the next six years, we would stop climate change in its tracks. The addition of those wonderful molecular machines, which pull carbon from our atmosphere, fix it in wood and bubble out oxygen in return, would halt the rise in global temperature and return it to a manageable level. Three hundred million years ago, trees took an environment with a toxic load of carbon dioxide and turned it into something that could sustain human life. They can do it again.

That simple idea isn't the end of things, though. What if you don't have the space or means to plant six trees? As I learned alongside Pat Lisheens and again when I formed the idea of the bioplan in my brain, you take the first step that you're able to take and you have faith.

A personal bioplan can take a form as humble as a pot on the balcony of an urban high-rise. One beneficial plant, a mint for example, releases aerosols that open up your airways. That plant does the same thing for the birds and other small creatures, and for the people you love and keep close. The true goal of the global bioplan is for every person to create and protect the healthiest environment they can for themselves, their families, the birds, insects and wildlife. That personal bioplan then gets stitched to their neighbours', expanding outward exponentially. If we each start with something as small as an acorn and nurture it into an oak, a master tree that we have grown and protect and are the steward of, if we have that kind of thinking on a mass scale, then the planet is no longer in jeopardy from our greed. We've become the guardians of it. It's a dream of trying to get a better world for every living thing.

Of course, we also need to protect all the forest we already have. It's all well and good to each plant a tree a year, but if we're clear-cutting the Amazon and ravaging the boreal at the same time, the positive effects are greatly diminished. The mathematics of carbon sequestration and release are a compelling enough argument to stop the destruction, but there's another significant argument in the idea of the bioplan. When we cut down a forest, we only understand a small portion of what we're choosing to destroy.

Though I have answered a great many questions over the course of my life in science, the list of things I hope to know about the natural world is longer now than it was when I left the University of Ottawa. Every medicine I've discovered within bark, leaf and root, or freely offered from a plant into the air, has hinted at the presence of untold others. Every unseen or unlikely connection between the natural world and human survival has assured me that we have very little grasp of all that we depend on for our lives. We still can't explain how water gets to the top of a tree—how the plant defies physics and causes water to run uphill. With such fundamentals still eluding our understanding, how can we cut down a forest? Just imagine the arrogance and greed of that— and the short-sightedness.

Here, in this protective effort, are even more avenues by which we can fight climate change. On a larger scale, we can band together to take on government and industry; we can keep informed of plans to destroy forests and fight them at every turn. I have been involved in many such efforts. Even when battling against multinationals, international organizations and governments, we have won. I have, for example, come on board to help save a large tract of mature forest in Pictou County, protecting the area that leads to and lines the MacKay Brook of the River John on the east coast of Canada with a group called Friends of the Red Tail. I have also written the master bioplan for the city of Ottawa, and served as scientific adviser for many environmental organizations both at home and abroad.

On a smaller scale, we can take on the role of guardian and steward within our own neighbourhoods and towns, as has been done to great effect in Winnipeg, Manitoba. The people of that city have come together to protect their elm tree, *Ulmus americana*, replacing yellow insect bands that trap the elm bark beetle, *Hylurgopinus rufipes*. In its asexual stage, the bark beetle spreads the deadly fungus *Ophiostoma ulmi*, an aggressive, pathogenic strain of fungus we are trying to curtail. These efforts have inspired others to do even more; they are sparks that set a fire of reawakening. If you have a large tree on your street, make sure your local council knows that you value it. Every opportunity to vote is an opportunity to put someone who cares about forests in a position of greater power and authority.

We are just shy of eight billion people on Earth at the time of this writing. With the forest we already have protected, we would have to add approximately forty-eight billion trees—my tally of six each—to absorb enough tonnage of carbon dioxide out of the atmosphere to halt climate change. Forty-eight billion may seem like a number too high to reach, but there's a simple way to get there: just take the first step and keep on going.

The Mother Tree

IMAGINE YOU'RE WALKING through a field. The sun is out, but it's not too hot. The ground is soft underfoot, a mix of meadow grasses, wildflowers, ferns and other plants. Every so often you cross a small patch of bare soil and the rich smell of earth rises to your nostrils. Once, as you step over just such a patch, an acorn falls from your pocket out onto the ground.

Fate smiles on this lost nut. Its shell and cupule part ways and the seed inside finds a welcoming home. Over days and then weeks, perhaps months, it splits its testa and pushes its single green shoot up out of the soil, it sprouts two thick leaves and soaks up sunlight and carbon dioxide, turning them into energy and fibre. It grows and grows—from seedling to sapling to mature tree. It reaches high into the sky and spreads, the tallest living thing around for some distance.

That height invites birds on migration, offering welcome rest and respite on their north-south, south-north treks. These visitors have feathers coated in an oil that contains the reduced form of vitamin D. As the sun hits this oil, it irradiates the molecule, breaking the second bond and turning it into a full form of active vitamin D. Perched on the now-mature oak, the birds preen their feathers, ingesting the vitamin, which helps them fight off disease and produce a greater number of viable eggs. In the process, they dislodge seeds nestled in their plumage and caught in their feet. They excrete a few in their droppings, as well.

The seeds fall to earth on the skirt of the oak and a certain number, like the acorn years and years ago, sprout. The fallen leaves of the tree, over time, have created a humus soil, rich in humic acid, which helps the roots of the new seedlings absorb water and nutrients. When it is strong and able, the oak even passes carbon and hydrogen in the form of food through its roots to the plants around it, particularly its own offspring, in an exchange akin to breastfeeding. Its offerings benefit only select species, but the list is long. In addition to other trees, ferns, lichens and mosses also benefit from the oak's generosity. Many trees thrive in the shade, waiting generations to grow. Each of those species is responsible for sustaining forty different species of insect. The oak is a metropolis and in the three hundred years it has taken it to reach maturity, it has also given rise to fresh, virgin forest.

In short, embedded in the DNA of the tree is the ability to create the specific conditions necessary to give rise to a rainbow of species. Here, given time, is the bioplan for an entire forest contained in the genetic material of a single seed. Of course, I couldn't watch this timeline play out first-hand. To understand how a tree sustains, and sometimes gives rise to, the forest around it, I had to work backwards from the forest itself.

———

In 1995, I decided that I wanted to do something to mark the close of the millennium that would be good for Canada and would encourage a greater public consciousness of trees. That summer, Christian and I drove to a plot of land we have on Prince Edward Island with Erika and a friend of hers, Laura, along for the ride. One stretch of the trip was a two-and-a-half-hour run across central New Brunswick, from Grand Falls to Miramichi. The highway is lined with trees, and the whole time you feel as if you're passing on the lone road through a massive tract of virgin forest. In New Brunswick, though, the forestry industry has a long history and a firm hold. The virgin forest is an illusion.

For the whole drive, Laura was marvelling at the trees. Though her excitement was beautiful, it was based on a lie. So I said to her, "Laura, you can't tell the book by the cover, girl. Let's get out of the car and take a walk."

We stopped and piled out. We headed into the treeline and walked for about a hundred feet and then saw what lay beyond the highway's buffer: a clear-cut landscape as far as the eye could see. It was like walking on the surface of the moon.

The devastation of that discovery playing out across Laura's face reawakened my own pain and anger. I decided right there that getting a good genome back out into the forests of Canada was a project worthy of a new millennium. If I succeeded, the results could last forever.

What we eventually dubbed the Millennium Project is, to my knowledge, the biggest tree-planting project ever undertaken in North America. Christian and I sent out 750,000 seeds and saplings to 4,500 recipients. The trees and seeds, all of them with dormancy broken, went out by mail over a period of years, each with documentation, provenance and planting instructions; twenty-two different rare species, all native, all propagated on the farm. We sent a catalpa to the Yukon that I received annual updates on for a number of years. It grew every summer, but only as high as the snow was deep in their garden—lending it an appearance more bush than tree. This is a common phenomenon in the northern latitudes, one that helps trees survive.

Among our goals with the Millennium Project was to send out trees with the best genetic material we could find. Christian and I had had our eyes out for those trees from the moment we bought the farm, but the project sharpened our sense of purpose. We increased the frequency of our expeditions and, gradually, an understanding dawned on me.

Every time we located a truly spectacular tree, the environment immediately around it was healthy and there was a feeling that everything in that zone of health was leaning in towards the tree. It was usually quite a large tree, and the ground around it offered up a smell of earth, fertility and vigour. Lichens would be growing on its bark and on that of the trees around it; this is an excellent indicator of atmospheric health, since lichens will not grow in polluted air. Different plants would dot the skirt of the tree, depending on the season. In spring, bulbs, corms and tubers, together with tough evergreen plants like the round-lobed hepatica, *Hepatica americana*, and various trilliums, sheltered from the colder temperatures by their blanket of leaves, signalled good health in the soil. Perennials, annuals and biennials would follow one after the other through the summer, and then fall would bring a huge fruiting of mycelium into mushrooms. Through the growing season, there would be a general bustle of animal and insect activity. A huge volume of butterflies were entering and exiting the environment around the tree. Holes in the ground indicated the presence of mice and other small mammals. The evening air held the clicking of bats flying into the darkness.

As I sat and watched these trees for extended periods of time, sometimes returning day after day for a matter of weeks, it was obvious that they were focal points of activity and vitality. They were epicentres of life in the forest, so I started referring to them as "epicentre trees." Later, when I'd learned more about the role they play in their environment, I changed my terminology. I now call them "mother trees."

Mother trees are dominant trees within any forest system. They are the trees that, when mature, serve up the twenty-two essential amino acids, the three essential fatty acids, the vegetable proteins and the complex sugars, be they singular or in a polymeric form of complexity, that feed the natural world. This menu protects the ability for all of nature to propagate, from the world of insects to the pollinators, to birds, to the small and larger mammals.

Many mother trees protect their home ground by producing an arsenal of natural allelochemicals that, beginning in the spring, flow automatically into the soil. This enables the tree to groom its own soil for the minerals it needs for health. The mature mother tree infuses the air around the canopy with aerosols, some as an invitation, others for protection. Mother trees can feed and protect other trees within the expanse of their canopy. They are the leaders of the community we call forests. And across the globe, forests represent life.

Mother trees have an effect on the oceans as well, as Dr. Katsuhiko Matsunaga and his team in Japan had confirmed. The leaves, when they fall in the autumn, contain a very large, complex acid called fulvic acid. When the leaves decompose, the fulvic acid dissolves into the moisture of the soil, enabling the acid to pick up iron. This process is called chelation. The heavy, iron-containing fulvic acid is now ready to travel, leaving the home ground of the mother tree and heading for the ocean. In the ocean it drops the iron. Hungry algae, like phytoplankton, eat it, then grow and divide; they need iron to activate a body-building enzyme called nitrogenase. This set of relationships is the feeding foundation of the ocean. This is what feeds the fish and keeps the mammals of the sea, like the whale and the otter, healthy.

In addition, mother trees produce pollen in the spring and begin to flush out aerosols into the atmosphere. As the aerosols rise with the warming air, they meet water vapour and blend with it. This is the cradle of the creation of our weather patterns. The human family thrives on a plentiful supply of rainwater—all from the bounty of mother trees.

I first observed this organizational structure in northern forests, but soon found out that all forests are based this way. The Brazil nut, *Bertholletia excelsa*, occupies the same position in the Amazon, providing the core around which all other species are healthy. There is a species of hickory, *Carya sinensis*, that mothered the forests of China. Mother trees are always nut-, legume- or acorn-producing species, because those sources of primary protein attract all manner of animal life. Mother trees are a common trait shared by every forest on Earth.

Of course, mother trees were the support structure for the ancient forests of Ireland, too. The Druids knew all about them. When Nellie spoke to the lone ash in Lisheens, she spoke as mother to mother, and traces of that understanding filtered down to me. So much of the knowledge I've gained in my scientific career was already there in Lisheens, in one form or another, but in Ireland there were no trees to help keep true understanding alive. Coming to North America, I had the trees, a great boon that allowed me to put what I'd been taught into context.

The genetic information of a mother tree is perhaps the most important living library there is. I already felt the truth of that in the presence of the trees we sought out for the Millennium Project, but it was back in Ireland, years later, that it really crystallized for me.

I was in County Clare visiting what remained of the ancient forest that once belonged to the last High King of Ireland, a single tree, the

Brian Boru oak. A *Quercus robur* perhaps 1,500 years old, the tree has command of the entire landscape as you approach it. It's said that the High King Brian Boru once shaded a thousand of his men atop a thousand horses under its canopy. Sitting under it a thousand years later, my mind reeled at the sheer size of the oak. I knew without a doubt that it was the mother tree for the area.

Corduroy roads once crisscrossed Ireland, and we know from dendrological studies of the logs they were composed of that trees like Brian Boru's were once common on the island. They were lost in the clear-cutting that eradicated the ancient forests of the Druids, all but that one tree. With them went the knowledge gained from thousands upon thousands of years of life. The knowledge of how to build virgin forest from a single acorn. The knowledge that we might need to save our lives and our planet.

In the boreal of northern Canada, the North American equivalent of that lost natural library still exists. Protecting it for future generations would soon become my life's next great mission.

Philanthropy
of the Mind

IN 2003, FOLLOWING the publication of *Arboretum America*, I was asked up to Lakehead University in Thunder Bay, Ontario, to be the keynote speaker at a government-funded conference on Aboriginal forestry and alternative uses for the forest. The idea, as it was first explained to me, was to come together to highlight the many things a forest could provide humanity beyond a towering stack of two-by-fours. Representatives of the Indigenous peoples of the northern regions across Canada were in attendance, along with academics, people from most of the major timber companies and others from organizations such as Greenpeace and the Sierra Club. Clearly there was a mix of groups and intentions involved, but the overall aim was plain: to highlight new, or less-explored, ways to profit from the forest.

A forest is a sacred place. I believe that with all my being. I have felt the proof of it, and so, I'd argue, has anyone who has set foot in one with an open heart. But knowing that it is a sacred place doesn't mean I'm closed off to the idea of people benefiting from the forest, even financially. We all need wood, in our paper, in our houses, in our lives. The medicines available in the forest are the second most valuable gift that nature offers us; the oxygen available there is the first. We don't need to be afraid to use that bounty, but we do need to understand it enough to respect it for the miracle it is.

A few hours before I was to give my speech, I had coffee with a medicine woman from the Rainy River First Nations. She told me her story. At seventeen, after the birth of her first and only child, she had been forcibly sterilized. She sat staring at her cold cup of coffee, with a film forming in a circle on its surface. Tears streamed down her worn face, though she did not seem to know it. They collected around her lips and chin. She was so traumatized that she was frozen in place with the memory of what was done to her and simply could not wipe them away. She, alone, grounded me with her courage. So much culture has been lost to the Indigenous peoples of Canada, especially those of the north.

In front of a huge audience, I took the stage. The image of the medicine woman I had been speaking with stood firm in my head, so I addressed the First Nations before the academics, timber boys and greens. I began, "There were two sacred trees given to the Aboriginal peoples of North America by the legendary Winabojo. These were the birch and the cedar. Great sheets of bark were gathered from the living birch trees in June or early July. The bark was stripped off, collected and stored. This bark was needed for domestic items like cups and kettles. It was heated and stretched to make serving dishes, boxes, coffins, wigwams, makuks for storage, cooking utensils, funnels and cones, meat bags, fans, torches, candles and tinder, dolls, sleds and your famous birchbark canoes. Once upon a time, your culture echoed throughout the world."

I went on to discuss the medicines of birch, including its uses for surgery and the setting of bones using look-alike limb braces strengthened with birchwood and stabilizers made from basswood twine. As I spoke, my words were simultaneously translated into the languages of the north—hundreds at once. The sounds of the unique Indigenous words fell around me like a waterfall. The medicine men and women who lined the back of the hall sat up; the timber boys were all ears. I ended, "Ladies and gentlemen of the north, remember Winabojo. This is your culture."

I was applauded off the stage and made my way up to the back of the hall, where the medicine men and women had saved a seat for me. The conference was over, save for a short talk from a representative from the United Nations. I parked myself next to a medicine man from Hudson Bay, a great bear of a person, and waited for the diplomat.

The spokesman there to represent the United Nations crossed to the podium and began to speak. I'll confess I was more occupied with people-watching than listening to him until a declaration, which he delivered in the same bureaucratic monotone, jumped out at me. I poked an elbow into the side of the giant from Hudson Bay and asked: "Did I hear right? Did he say they're planning to cut down 50 percent of the boreal forest?"

My companion grimly confirmed I'd heard correctly with a nod of his head and one terse word, "Yup."

Here I'd just delivered a talk on the medicines of the forest and how more money could be made by not cutting down all the trees, and this fellow was talking about levelling 50 percent of the northern boreal, a forest the size of the Amazon that had taken thirty thousand years to grow into its unique and irreplaceable form. It's a forest that maintains the salt-water conveyer belts of warm and cold water in the oceans. It is upon this flow that the seasonal weather patterns of the globe depend.

The plan he was outlining was genocide as far as I could tell. *To hell with this,* I thought. *I'm going up there.*

I walked back down to the front of the hall and onto the stage, grabbing the microphone from the UN representative. "This is genocide you're talking about," I told the crowd. "All you people here come from the north. This plan would diminish your existence to the point you'd die. It's also genocide for all migratory animals—the birds, fish and sea mammals. The boreal forest system is the workhorse of the world. It cannot be replaced. Opening up the boreal will expose the benthic phenols that lace the waterways and lakes and carpet the underbelly of the tundra permafrost. These phenolic graves of plant remains—the layer of lakes and rivers and swamps where organic material decomposes under the pressure of water—must stay frozen; otherwise, the atmosphere will be filled with carbon dioxide and methane from that decomposition."

I went on to suggest that the peoples of the boreal are the real stewards of the planet. As such, I said, they should be paid for their duty of care for the rest of us. I also suggested that the World Bank step in with a loan to back this idea. If the boreal forest system remains intact, it gives the planet a fighting chance that we can all combat climate change.

I was so hopping mad that I didn't even notice, as I spoke, that the medicine men and women in attendance from all across North America had made their way up onto the stage to stand behind me in a large, silent semi-circle. When I finished speaking, a medicine woman came and took the microphone from me and said, "Diana speaks with one voice with us."

The silence in the hall after she said that was explosive, almost as if everyone in attendance had stopped breathing. The organizers were furious with me, I could see it in their faces. I carried the fire of that talk with me all the way home. As soon as I stepped off the plane, I said to Christian, "Goddammit, we're writing another book. Get your camera ready. We're going up into the boreal."

When the conference proceedings were later published, the only talk left out was mine, which had been the keynote. I was expunged, but the Indigenous people who were there remember my talk to this day.

———

That northern Ontario forestry conference wasn't the spark that lit my activism. From my childhood in Lisheens, I'd been taught to freely share anything I was able to, especially my knowledge, and to always look for ways to improve the world around me. I'd never had the money to engage in the type of philanthropy favoured by the Marias of the world. Instead, I gave back through something I came to call philanthropy of the mind, using my scientific knowledge and all the energy I could muster to advocate for any worthy cause that came through my door and to spread awareness of the issues nearest to my heart.

In the early 1990s, I had propagated and sold hellebores from Bosnia to raise money for women affected by the war in the former Yugoslavia. That effort paid for fifteen thousand surgeries, bandages and electrocardiogram machines for Médecins Sans Frontières and a safe house in the city of Tuzla. Before that I had sat on the parks board in Merrickville and fought to legislate chemical-free byways around farmers' fields so there'd be some wild territory for pollinators. Christian and I had fought to amend a new section added to Prince Edward Island's environmental protection act to increase the protection of riparian areas as a safe ground for such at-risk species as the long-billed curlew, *Numenius americanus.* I'd engaged in a two-year communication with the federal health minister about the potential, and as yet untested, health threat of genetic modification. The Millennium Project was also an activist effort. I'd worked on behalf of schools, libraries and churches all over Ontario, and at every talk I'd ever given at a horticultural society, over more than four decades of engagements, I brought the list of endangered trees and plants for the particular area where I was speaking, gave the information to the crowd and told them, "You are responsible for these trees."

Even though efforts to make the world a better place weren't new to me, that forestry conference did crystallize my thoughts and actions in a new way. In alerting me to a grave threat to one of the most important natural wonders on Earth, it widened my viewpoint, allowing me to see for the first time that I could take on much bigger challenges than I had imagined. In fact, I actually needed to take them on. Christian and I spent the next several years working on our new book, which was meant to raise awareness of the boreal.

I was still fighting for that recognition of this vital forest system that rings the planet in 2010 when we published our book on the northern forest, *Arboretum Borealis: A Lifeline of the Planet,* as well as my collection of essays called *The Global Forest.* That collection came to the attention of the CBC Radio program *The Current,* and its host, Anna Maria Tremonti, did a feature interview with me. Filmmakers had been interested in my work for a while, but I'd always refused their overtures. That day, a documentary maker from Winnipeg, Jeff McKay, heard me on the radio and stopped his car to listen. Then he drove to his studio on Princess Street to talk to his team about contacting me to discuss the possibility of doing a forest film based on my work. As far as I was concerned, he was a bit of a wild man, totally unlike the other filmmakers who had approached me. I thought, here was a man who would go the distance with me. It was also important to me that he was Canadian and that any film we made would be shown here, in my new home. I ended up writing the director's script of the movie that became *Call of the Forest: The Forgotten Wisdom of Trees*. We travelled all over the world for the filming, a process that took five years. I also worked for a year on the tree-planting app, *calloftheforest.ca*, that went along with the film, which keyed species to plant to the regions where people lived, along with information on the medicines of the trees.

In the early stages of filming, a reporter named Alexandra Paul wrote a piece on me for the *Winnipeg Free Press*. Alexandra was friends with the environmental activist Sophia Rabliauskas, a leader of the Poplar River First Nation who, about a decade earlier, had secured protected status for two million acres of virgin boreal forest on the eastern side of Lake Winnipeg. Alexandra told me that Sophia would be near my home visiting family in the coming weeks and wanted to stop in at my place for tea and a chat. I agreed and Sophia turned up with some friends in minus-thirty-five-degree weather later that month. We talked for more than five hours.

During that first session, Sophia told me about Pimachiowin Aki, a massive area of virgin boreal forest on the Manitoba-Ontario border. The name, she said, means "the land that gives life" in Ojibwe; clearly these were people I could see eye-to-eye with. For thousands of years— the entirety of their recorded history—the Poplar River First Nation had safeguarded that forest, and in the years since they had secured protected status for that first two million acres, they had been fighting to have the whole of Pimachiowin Aki declared a UNESCO World Heritage Site. They had already filed an application with the UNESCO office in

Paris and were waiting for a response. I had not heard of the area or their effort before, and I told Sophia I would do anything I could to help. If I visited and saw the forest with my own eyes, I said, I could write a scientific paper on its value in support of their cause.

We parted on happy terms. They were clearly making a serious and wholehearted effort; I believed in them. So I was stunned when I later heard from Alexandra that the UNESCO application had been rejected. In retrospect, I shouldn't have been so shocked. They were applying to a body overseen by the very organization whose representative had so upset me years earlier with his pledge to cut half the boreal. And humanity's indifference to nature had stopped being a surprise long ago.

I knew, though, that they were far from beaten, and that now is the perfect time for this particular fight—the fight for nature. Yes, we are closer than ever to climate catastrophe. Yes, plenty of people on the planet are still indifferent or even defiant in the face of that fact—and often paid to be defiant by vested interests. But we are also lucky enough to live in a time when more people than ever care deeply about the natural world, understand that the forest is an ancient, sacred place, and want to do something to stop our suicidal pursuit of profit and "progress."

I had the great good fortune to be born just in time to receive my Celtic education. But in the years between my graduation and my departure for North America, I watched my teachers pass on and the embers of that tradition fade away until I was left alone with them. I took that spiritual view of nature with me into academia and found that it wasn't welcome there; I was told that "science and the sacred do not mix." Among academics, a scientist was expected to know better than to trust the knowledge of Indigenous woodland cultures. These attitudes and more drove me from the institutions of science and education onto the margins, and I worked there for many years before finding a significant audience with whom to share what I'd learned, what I'd always known and what I had to say. Now, though, I know that belief in the spiritual and scientific value of forests doesn't have to remain at the fringes of our culture. It can be a mass movement; the interest is there, the people are ready—the youth are afraid, the children anxious. And it's not too late to act. For all of that, we should be deeply grateful.

We arranged with Sophia to do some filming for *Call of the Forest* in Pimachiowin Aki. I flew with our film crew to Winnipeg and then travelled north to the lake that shares the city's name. There we had to cross from the west shore to a point on the east side at the mouth of the Bloodvein River. Lake Winnipeg is very shallow, so the wind whips up very high waves. It can only be safely crossed when the conditions are right. The Chief of the Bloodvein First Nation, William Young, picked us up in a big boat. When we boarded, I noticed golden eagles circling above. And as we glided over the surface of the lake, they followed us in a convoy, some to our right and others to our left. When seen from right below, these birds look massive.

We arrived at the lodge where we were being put up during our stay and unloaded our gear. It was at the junction of the river and the lake. The water of the Bloodvein was clear, pristine and beautiful—safe to drink. It's sad that's rare enough that I need to point it out. Bald eagles joined the golden eagles circling and swooping above. They would occasionally plunge down to take what they wanted from below. The river was packed with fish, beaver and otters, whose kits played around my feet as we filmed. The air was pristine, fresh, wild—as effervescent when you breathed it in as champagne bubbles on the tongue.

On both sides of the river stood forest that has been untouched and actively protected by the Indigenous peoples for five thousand years. It is the truest embodiment of virgin northern boreal forest on Earth, and its biodiversity was unbelievable. Great carpets of lichen covered the forest floor. Maybe a foot thick, they made a crunching sound as you walked across them. The trees were old and the lichens crept up them, doubling the forest's chemical capability to purify the atmosphere. The owls were out in great numbers. And the place was rife with the noises of large animals, maybe even bears.

That night I couldn't sleep. I was excited, but that wasn't what kept me up, but rather the sense of an unseen presence in the lodge. When I looked at the clock and saw it was six in the morning, I figured it wasn't worth staying in bed any longer. I decided to get up and take a walk down to the shore to see the dawn mist rising off the river.

I got dressed and went outside. The sun was up, but just barely. Immediately a great flock of birds darkened the sky. The click of their wings was like a thousand knitting needles. They descended in a rolling cloud on the water meadow in front of me. As they lighted, I saw they were red-winged blackbirds. To concentrate in such numbers on a single area, I knew they had to be eating something. I went to the edge of the water for a better look at the plant they'd descended on and saw that it was a wild rice. "They're feeding on wild rice!" I called out, hoping someone would wake up to witness this with me. "They're feeding on wild rice!"

I went charging back to the lodge to get the sound engineer, Norman Dugas, so he could record the birds for his three-dimensional sound project. It took him half an hour to wake up and get his gear to the dock. By that time half the birds were gone. There were canoes by the water and I got him to take me across to where the birds had landed. I wanted a fully intact specimen of the rice for dissection. My captain dug a plant out with a canoe paddle and I rode back holding it with my arms extended across the canoe to keep it safe. I wanted to do a full botanical dissection on this plant in the lodge, documented with photographs and notes.

It was indeed a wild rice of the genus *Zizania* and there was nothing else like it available in North America. It was seven feet tall, yet it hadn't been the tallest plant in the water meadow. It had an articulated head that I knew was capable of tracking the sun's path across the sky in a rotary motion. It has been a food crop of the Indigenous peoples in that area for five thousand years, never crossed or hybridized. It was a treasure. I identified it as a variant of wild rice by the name of *Zizania aquatica* var. *angustifolia*. A true find for science.

I used the existence of that species to plead for the area's importance. I argued that it was a natural treasure and a living monument to the culture of the Ojibwe people, who had protected it for millennia. I also described in detail the significance of the forest in Pimachiowin Aki. The patterns of growth and replacement in virgin forests of the boreal north have not been scientifically studied in full detail. The presence, for instance, of myxomycetes, or slime moulds, growing in conjunction with lichens have not been recorded in such a formation as I witnessed there. Slime moulds are a primitive species that have become important in medicine because they appear to use the call command of computation, like the opening and closing of stoma on a leaf or the way T-cells respond to the chemical call of the immune system. Slime moulds are doing this in response to chemistry and not the mechanical push of a button on a computer keyboard. Slime moulds occur and move together by some unknown means to survive the oncoming cold of winter. Their mass movement is still a mystery to science.

My plea was added to the existing efforts of Sophia and the First Nations of the Bloodvein and Poplar Rivers. In August 2018, Pimachiowin Aki was named a UNESCO World Heritage Site, one of very few on Earth given the designation for both natural and cultural significance. Sophia's efforts culminated in eleven thousand square miles—an area the size of Denmark—of that almost unimaginably beautiful boreal forest earning protected status. Pimachiowin Aki is now the largest protected land mass ever.

You've got to stay on your hind legs and take a swing—all of us do. You've got to take a first step towards a goal that seems unachievable, and have the integrity and courage to believe that you will reach it one day. We all have immense courage. Every one of us is capable of extraordinary things. Pimachiowin Aki is living proof of what can be accomplished when we believe in ourselves and keep moving towards the impossible.

———

Stopping climate change in its tracks can feel like the impossible. The latest science tells us that as of 2019 we have just ten years to halt global temperature rise. If we leave it any longer, it will be too late. Instability in the natural world will generate chaos in human institutions, and we will have squandered the immense gifts that trees have given us. But my life and work have taught me that nothing is ever as dire or insurmountable as it seems, and that the natural world's powers of regeneration and restoration stretch far beyond our understanding.

In these pages, I have laid out a plan for every one of us to fight for and save the global forest, our planet and ourselves. And it's not a complicated plan. It's as simple as protecting what we still have and planting one native tree each per year for six years. We can achieve that goal and we have to do so—not just for ourselves, but for all the children who will follow us, the future generations on whose behalf I've worked all my life. And when we achieve that goal, the rewards will extend far beyond the security of a healthy, stable climate.

There is a deity in nature that we all understand. When you walk into a forest—great or small—you enter it in one state and emerge from it calmer. You have that cathedral feeling and you're never the same again. You come out of there and you know something big has happened to you. Science allows us to explain a part of that sacred experience. We now know that the alpha- and beta-pinenes produced by the forest actually do uplift your mood and affect your brain through your immune system. That pinene released by the trees into the air is absorbed into your body. It tightens you as a unit and makes you reverent towards what you're seeing. Simply walking into a forest is a holiday for your mind and soul, allowing your imagination and creativity to bloom. I think that is a miracle, and there are so many other miracles of the natural world left for us to discover.

We will feel the joy of those miracles. We will save the forests and our planet. The trees are telling us how to do just that—all we have to do is listen and remember.

PART II

THE CELTIC
ALPHABET OF TREES

I HAVE ONE more gift from my Brehon wardship and my life in science to convey to you in this book: my annotation of the Ogham script. This is the first alphabet of Europe, in which every letter is named for a tree or an important companion plant of trees. According to the ancient Celts, the song of the universe dictated this script to a young man called Ogma. It was a treasure of the woodland culture of the Celts and especially the Irish. The famous *Book of Kells* came into being because of Ogham. This alphabet gave birth to the literacy of Ireland; words like "car" and "hospital" that are still common today come to us from that ancient language. What is really remarkable about Ogham, though, is that the philosophy behind the language carries a way of thinking about the rebirth of the forest, namely, to consider how intimate the connection is between forests and humans.

Before starting the alphabet, I want to ask you to do a simple experiment for me. After a dark, overcast winter's day, or a prolonged period of rain, go out into the sunshine. Take a stand and spread out your arms, palms up. Tilt your head up, too, and let the sunshine land on your face, your hands, the rest of you. Feel the sun on the surface of your skin. With this act, you are becoming like a tree. You are acting like a tree, with your arms spread out towards the sunshine just like the leaves in a canopy.

In Lisheens as a young girl, I was asked to become a tree in the presence of sunshine. The feeling you have on your skin is a dance with the short-wavelength energy of the sun. This dance has a name in the ancient world of the Celts. It is called the song of the universe, *Ceolta na Cruinne*. It is real. You can feel it for yourself. It is the song heard by the ancient Celts before young Ogma created this alphabet, which stitched together song and story, medicine and faith, people and the forest.

A
Pine
Ailm

The giants that lined the waterways of the Celtic world were pines. The golden bark of these conifers matched the hue of the sunset; their plumage of stark green needles plucked at the cobweb clouds of the sky. Weary with age, they gently sprinkled their needles to the ground, where the soil soaked up the bitter acid of the decomposing leaves.

These pines had smaller companions, now rare. They were deciduous trees with a reddish bark called strawberry trees, *Arbutus unedo*, that sat in the piney shade. Both were found together at the shoreline, where their fallen leaves wove into the humic soil. The strawberry tree carried clutches of creamy bells in the autumnal months that ripened into scarlet, warty berries. They were the choice heathland fruit of the Druidic physicians, known also to the Arabs and the Greeks.

The Scots pine, *Pinus sylvestris*, with its soft antiseptic wood, was the darling of the Celtic kitchen. Butter and milk was safe in its keeping, for its wood kept rancidity away. The light weight of the wooden domestic objects made from the Scots pine were always welcome to weaker wrists. This pinewood was easy to clean and its smell was always fresh. The lady of the house, *bean an tí*, had her own name for this wood—*déil*. It was also the wood of the potter's wheel.

In the oral history of the Celts, their pine bore a name millennia before the birth of Christianity. The people used groves of pines at the seashore and along lakes and rivers as visual markers for navigation. Their highways were waterways, along which they paddled their small boats, *curach*, or rowed their bigger vessels, *bád*, from point to point with a minimum of effort. The fresh, fragrant foliage of the pine was always waiting for them, both in language and memory.

When memory stepped into language, a new idea was born for the Celtic peoples—the art of writing. The thought of such a thing as a written form of communication had been swirling around them, in the Sanskrit carried from India. Writing was the next step for the extraordinary oral culture of the *Keltoi*. The printed word is a platform for change and evolution. The possibility of the written word drenched the Celts like a torrent of rain, soaking the dry turf of poetry in Ireland so it sent forth shoots.

Legend says that when a young man named Ogma created the Ogham script, nature came calling, because all imagination—even scientific imagination—originates in nature. Young Ogma looked around his neighbourhood and his eye fell on the Druids' sacred life forms—the ancient forests.

So the alphabet of the forest was created. This new writing held the philosophy of the forest and of the millennia of oral culture that went with it. A written word is no small thing; it holds a sanctuary of thought. It is a record of renewal from the culture that originated it. With it, the Celts created the second-oldest written language of the western world.

First Ogham was written on long rectangles of the outer bark of aspen and hazel, harvested in spring. Filled with fluids, this piece of spring bark naturally curled inwards and felt wet and slimy with protein-filled cells that are similar to caesin, a milk protein carrier for pigment in paint. The surface could be written on, and it held the writing in perfect condition as long as the bark remained dry and in good shape.

The second era of writing was on very large squared-off rectangular stones. They were placed in positions in the countryside almost as memory monuments, and perhaps they were. When I was a young child in Ireland, these stones were also used as markers for the ancient footpaths that crisscrossed the hills and mountains. The surviving Irish farming community were fully aware of the importance of these Ogham stones to their past. They represented a common heritage and were not interfered with in any way. Those fields with stones were used only to graze sheep or cows.

The stones were inscribed on their faces in the Ogham script using tools of iron and bronze. Metal work was a specialty of the Celts, who also indulged in design. Over millennia the script has weathered, but it can still be read by touch and finger tracing.

The Scots pine was extremely important to the Celts. Ogma himself must have named it *ailm*, to become the letter *A* in his alphabet of the sacred trees. In the Ogham alphabet, it is represented as a simple cross, a vertical line intersected by a horizontal one. The letter *A* in Irish has two sounds, one short and the other lengthened by an accent called a *fada*. This was a useful sound for the poets and bards.

The medicine of the Scots pine has been lost. What has come down to us in its place is a carefully guarded memory. We know the Druidic physicians considered their evergreen pine to be essential to health. They prescribed walking in pine forests to aid in breathing and to rid the lungs of the toxins of colds and flus. Their first recommendation for someone suffering from such an infection was a twenty-four-hour sweat wrapped in a linen sheet, after eating milk boiled with onions, wild if possible. When the patient's energy returned, they were asked to walk in a pine forest. This process of forest bathing has recently been tested by clinical medicine and confirmed to be beneficial by patient trials.

The pine bears a canopy of needled leaves that, as the temperature rises, produce an atmospheric aerosol complex of a biochemical called pinene. As they become airborne, the pinenes enjoy a levorotatory twist. (Levorotation is like a kite that can fly only to the left.) This form of molecule is easily absorbed by the skin and the surfaces of the lungs, and has recently been shown to boost the human immune system. The beneficial effects of a twenty-minute pine forest walk will remain in the immune system's memory for about thirty days.

B
Birch
Beith

When I was a young student in Ireland, I once took the high road down into Bantry. My intention was to drift through the Killarney Lakes and examine what spring had on offer. I was armed with my magnifying loupe, collecting bags, a ham sandwich and a thermos of tea. A small, cozy, green valley caught my attention. In it a large Scots pine wandered close to a beautiful stand of gleaming white birch trees, eager for the sunshine. Something in that trove of trees called out to me.

I scrambled over the fence and braved the blackberries using my thermos as a buffer. I aimed, as the bird flies, for the one enormous boulder against which the Scots pine leaned drunkenly. Arriving at my stone boulder of a table, I was about to put down my thermos of tea when I noticed the green tablecloth. I stared at it in astonishment. I had discovered the cloud fern, *Hymenophyllum*, a first for Ireland.

In addition to the cloud fern with its tablecloth-thin layer of cells spreading over the boulder, I met Ireland's native birch for the first time. These trees, *Betula pubescens*, the common white birch, were just making a comeback, returning in odd spaces after a five-hundred-year exile. In the Celtic world, the birch was known as the lady tree and held in high esteem. Its bark wears a white periderm like glistening talcum powder. The periderm allows for reflection in daylight, but in moonlight, which itself is refracted light, the birch glows a silver white. This moonbeam reflection earned the tree its ancient name, *beith gheal*, gleaming birch. The reflection, like a formal evening dress, goes all the way to the ground.

The Celtic woman was respected. Love, *grá*, was showered on her. This *grá* was especially true on the home front. Trees were often named after women, as were places, wells and holidays, like Lady's Day in August. All Celtic women reaped the benefit of this regard, from the *Bean ríon*, queen or woman king, to the *bean luí*, or concubine. The escapades of the poaching pirate queen, Queen Medb of Connacht, were never lost on women or on men. This celebrated beauty led a vast navy into battle from the rim of the western Atlantic seaboard. Her plunder also included the odd gentleman who fell prey to her magnetic sexual appetite. The tales of Queen Medb vied with Homer's *Iliad* throughout the Celtic world, and still do today.

The lady tree, the birch, was baptised *beith* well before the birth of Christ. In the oral culture of the Celts, *beith* was a temple word, a trigger for the meaning of life—a three-way ply of body, mind and soul. The word *beith* means to exist as a mystical constant outside of time. There is an ancient Celtic saying connected to *beith* and a version of this saying is found in every major world religion: "*An te a bhí agus atá*," or "He who was and always will be." It is the seeding of the divine into the birth of mankind. The Irish birch carries a memory of this gift.

At the foundation of nature is the eternal question of life and its meaning. The *ollúna*—male and female scholars from the educated elite of the Druids—took these questions seriously. The philosophy of life was passed down successive generations of the endowed families. The words were tossed about, their meanings changed and turned inside out by riddles. The bards skimmed the milk of words into the cream of verse and song. The royal court of Tara solidified the cream into the simple butter of bardic rhyme so it would endure the test of time.

The Celts took their birch, *beith*, and gave it the letter *B* in the Ogham alphabet. The drawing of the letter begins with a vertical line. From the centre of this line, a second short perpendicular line is drawn to the right.

All of the seafaring nations of the north made extensive use of the birch, from the birchbark canoes of North America to the curach of the Celts.

The Druidic physicians, male and female, knew the medicine of the birch. A tisane made from pouring boiling well water on mature leaves is an ancient treatment for urinary tract infections. This mildly diuretic tea is thought to have a gentle antiseptic action on the delicate tissues of both the male and female urinary tract. This tea is still used by Indigenous peoples throughout the circular crown of the boreal forest system that sits below the Arctic.

Probably the most stunning aspect of the birch is to be found in modern times. A tree that was once known as part of the commons is now in the hands of the multinational corporate world. The pearl of wintergreen has been converted by steam distillation into a little white pill. A cool fourteen million of these "aspirin" pills cross the sales counters of drug stores each day. The pill is used for pain relief and reduces fevers and inflammation. It has also been found to be an anti-clotting agent that keeps blood flowing in the general circulation of the body.

But there are still more biochemical medicines to be found in the birch, one of which is a modern surprise. The birch produces a regulatory phytochemical in the dark hours of the night. This little gem, betulinic acid, turns out to be a growth regulator. It induces human melanoma cells to commit suicide. It is an anti-cancer drug that keeps the tree in a waiting pattern, controlling its growth until conditions of sunlight or moonlight improve.

The birch also contains a strange little sugar called xylitol that inhibits a bacterium responsible for tooth decay. This bacterium is also responsible for ear infections in young children. Now xylitol is being put into chewing gum to create healthier teeth.

My evergreen treasure trove is still growing in Killarney, the land of my ancestors. I would be willing to bet that the next time I visit the valley, some undiscovered gem will be waiting for me there. Nature is full of surprises.

C
Hazel
Coll

An ancient castle of the O'Sullivan clan, Ardea, is perched on a cliff in County Kerry near Kenmare. The O'Sullivans were pirates, named for their pirate king, *Súil Amháin*, One Eye. The castle sits on a rabbit warren of hidden chambers, some of which have passages that stretch over a mile. There was a rumour when I was young that Spanish gold bullion was to be found somewhere in that warren. The last of the O'Sullivan tribe, my cousin William, once decided that the treasure would be his.

William was a *collóir*, a dowser, who had recently learned that he was as sensitive to the presence of gold and silver as he was to his usual quarry of water. He arrived at the castle with a forest's worth of hazel, or *coll*, which he proceeded to cut into a series of wishbone shapes that reached his waist. I was there to witness that as he walked, the hazel dowsing rod, *slat choill*, almost twisted itself into a knot, the pull on his hands was so powerful. At the end of the day, William O'Sullivan learned that there was no gold or silver to be had, just a series of streams running under his castle.

The hazel tree has an unusual history. The last ice age forced the land to rest. Frost action, combined with the scouring of glaciers, pulled the riches of macronutrients from the surface of rocks. After the ice receded, the land was as fresh as a new pin. The low-growing hazel was perfectly suited to be the first crop of forest.

Scientists have taken core samples of the earth from periods just after the last ice age. These samples, everywhere, are contaminated with a huge amount of pollen from hazel woods that seemed to drift in shoals on the surface of the soil. Some of it would have come from climatic stress on the trees that forced them to reproduce; the other factors for that first efflorescence of pollen are unknown. We are now witnessing releases as bountiful as they were at the end of the last ice age. Make of that what you will.

The Celts loved to eat hazelnuts. Indeed they were a favourite table treat for Celts from Ireland and all across their civilization into Turkey. They were also enjoyed further afield in China and Japan. The Celtic farmer practised a sophisticated method of two-tier farming involving both hay and hazel. Hazel trees enjoy the stability of roots feeding undisturbed in a field of grass. A hedgerow of hazel protects the hay from blowdown. So both grew together to benefit each other—the better the hay crop, the larger the nutmeat. The hay was cut, dried and harvested first. The ripe hazelnuts were gathered immediately afterwards.

Hazel trees travelled with all of the old civilizations. The fresh bark of the hazel tree was removed by the Greeks and used to write on. The word passed along in language, today making an appearance in "protocol," from *protokollen*: the first sheet of papyrus roll from which a document of record is made.

Before the advent of drilling machines and drilling rigs with pounders or diamond bits, people depended on instinct alone to find water underground. Individuals who inherited this instinct for discovery were of great value to any community. The water for my well in Ontario was found by the local dowser, who also happened to be the well driller. After he estimated the greatest flow rate per minute and per hour, he drilled the well and the house went up over it.

The Canadian dowser did exactly as my cousin William did. He cut a piece of local hazel in the shape of a wishbone that reached his waist. He grasped each fork of the hazel firmly and then began to walk, pointing the *V* of the hazel to the ground. He crisscrossed a fairly large portion of ground and then he felt a twitch. Encouraged by this, he followed the twitch until the hazel rod became frantic. Then he pushed back his cap with a thumb and smiled. "Guess we've hit water! Be drilling here. It's a good 'un. Real good."

The hazel tree, *coll,* was called to membership in the pantheon of sacred trees, or *bile,* because of its food value and its strange power of divination for water. The tree was added to the alphabet of the Ogham script, where it carried the letter *C.* The written form of this *C* was a vertical line from which four parallel lines were drawn to the left. The letter *C* is much used in the spoken form of the ancient Irish language. It can change its sound with a dot on the *C* to a harder, sharper sound like Russian, a sound useful for emphasis in poetry and bardic rhythm.

Beside its value as a medicinal tree, the nutmeat of the hazel was always considered to be a beneficial health food by the Druidic physicians. Every person had a right to collect hazelnuts under the Brehon Laws, like water and air—a commonage of health equal in share, from the top to the bottom of society.

The medicines of the hazel tree were common knowledge all over the world, once upon a time. The mature leaves of the larger hazel, *Corylus colurna,* were dried and smoked in Asia. The Indigenous peoples of North America used their native hazel, the beaked hazel, *Corylus cornuta*, as an infant medicine to reduce the fevers of teething. They also used it as part of a complex medicine to treat loneliness.

In more recent decades, a powerful medicine has emerged from species of the hazel tree across the globe—a medical biochemical called paclitaxel. This is an anti-proliferation agent that halts the march of cancer cell growth and is currently in use in cancer treatments.

The ancient practice of medicine, ranging from Ireland across Europe to the Middle East and Asia and even North America, has something to tell us about the hazel. It is an exquisite expression of biodiversity: without the hazel, we would not have this big gun to kill cancers.

D
Oak
Dair

The oak is the darling of the Celtic world. As I discovered as a child, the peat bogs of Ireland occasionally spit out chunks of oak trees that are pickled in the brine of ages. The open wood grain of the oak darkens with the acids in the bog waters to a gem-like quality. This heavy black bog oak, rich with the scent of centuries, becomes a work of wonder in the hands of a sculptor.

The Gaelic name of the Irish oak, *dair*, related to the Greek and Sanskrit forms of the word, managed to elude redaction by the Penal Laws. The oaks themselves, *Quercus robur*, barely survived, though, usually hanging on only as specimen trees on Anglo-Irish estates. Much of the language was not so lucky. During the five hundred years of England's rule, the Gaelic language, with its unforgiving pronunciation, solid rules of grammar and exquisite assonance of pride and poetry, was put to the sword, too. But the English found *dair* easy to say, so that word survived.

The Irish oak, once grandstanding in magnificent groves, is the real *lus* or healing herb of the Druidic physicians. The oak can live to almost a thousand years and sometimes longer in the rich soil and temperate rainforest conditions of the garden of Ireland. As the tree matures, the main trunk produces horizontal limbs that bear cluster canopies of oak leaves at the tip. These, too, increase with age. The ancient forests of oak match canopy to canopy in a bid for the sun, which forces the tree to make decisions. It decides to lengthen the branches and tilt them down to the soil for support and rise again into an open space to receive sunlight. Where the branches touch the soil, a series of roots plunge into the ground as a secondary feeding mechanism for the growing tree.

On the branches of an ancient oak tree, the periderm and cortical tissues produce the fine dust of compost. This turns into soil. The damp soil found on this upper layer of the horizontal branches plays host to a series of rare fern and moss species that thrive in perfect conditions. This sets the stage for visits of the songbird, the mistletoe thrush, which flies in with sticky seeds. The plant parasite called mistletoe, *Viscum album*, takes hold. This is the magic herb, *drualus*, of the Druids.

The oak yields another ancient medicine. In old age the living oak endures wind effects on its canopy that add to its weight. The wind twists and turns the canopy, inducing torque on the trunk. A water appears. It was called *uisce dubh*, or black water, by the Druidic physicians. It is a mighty molecule, a polymer called a gallo-tannin, still in use today, especially in burn wards.

In the past the oak was a feeding tree for man and beast. Some of the six hundred species of this family grow edible acorns. Other acorns have to have the tannic acid extracted before they are sweet enough to be ground into flour or roasted. This was the basis of acorn-consuming cultures all over the world. You can still find large edible acorns available today in many Arabic and Asian food markets.

Botanically, the oak is the king of the plant world. Each tree is a metropolis for insects, butterflies and pollinators. The Indigenous peoples of North America used the oak as a thermometer for plant growth. The tree's longevity is remarkable, as is its attunement to the sun. The tree carries its own sun screen. The fallen leaves on the ground continue their photo activity and release a root-growth hormone for the acorns called abscisic acid.

The love affair between the Druids and the *dair* is the stuff of legend. Bound in the wood of the oak is a timeline of the millennia, an accurate account of the tides of time. The oak is a sacred tree, or *bile*, and its name, *dair*, was given to the letter *D* in the Ogham script. It is represented by a vertical line with two parallel horizontal lines to the left.

A Druidic legend claims that the tree is the beating heart of the planet and that a time will come when people will repair these sacred oak woods, beginning in County Clare, Ireland. The legend says that the idea will catch the world like a wildfire.

I remember meeting the Brian Boru oak for the first time in Ireland. This massive wooden altar to the gods lounges across a glacial hill spreading far and wide, like the lion king on the Serengeti plain. There is a confidence in this tree, matched to its absolute elegance, which commands the landscape.

The tree has to be circumnavigated like the globe. Its wonder expresses itself as you approach it and look up. Everywhere there is canopy overhead generating an atmosphere of its own. Only when you look down and across at it from a nearby hill do you realize that it is a single tree with a single massive trunk that can be compared quite easily with an iconic North American redwood.

The Brian Boru oak is guarded not by a dog but by a massive black bull with a huge iron ring in its nose. This creature hugs the darker shadows of the tree like a *síoga*, or fairy. One footstep towards the tree awakens the thunder of its hooves as it lowers its head and gives its all to the gallop. The bull and the oak are friends; one grows while the other groans.

This is the last great tree of the temperate rainforest of Ireland. It is one example of the ancient forests of Europe. It is also the one tree that defied my efforts to clone it.

E

Aspen
Eabha

Nature is filled with eccentricities. The Celts took this in their stride and were able to tabulate some rules out of the chaos of nature.

One guidepost in that chaos was the aspen tree. It was the weatherman of the Celtic world. People watched this tree day and night for signs of oncoming weather. Its egg-shaped green leaves are more transparent and thinner than the leaves of other deciduous trees. The feel of the mature leaf is like silk. This silken sail hangs on an extra long petiole, or leaf stem, which is attached to the tips of branches to form the aspen's canopy.

In addition, many of the leaves have a twin glandular system on the leaf end of the petiole that acts as a miniature counterweight, somewhat like the pendulum of a grandfather clock. The surface of the upper leaf is smooth and the lower surface carries microscopic hairs that are matched to the uneven edges of the leaf's plane. As a result, aspen leaves catch the wind and flutter. Even the gentlest breezes cause them to move.

The aspen's name in science is *Populus tremuloides*, or the tree that trembles. In languages all over the global forest, this name holds for the respective resident native species. If the wind catches the leaves to make a rustling noise during the night, there will be rain the following day. If the wind lifts and turns over the leaves, showing their whitish lower side, then a gale can be expected. A dry rattle of the leaves is a sign of impending showers.

The Celts took this knowledge of the trembling leaves to the blackboard of the night skies for closer inspection. If there is a halo around the moon, transient weather is on the way, from dry conditions to wet weather. If the count of stars within the halo is one, two or three, this represents the number of days until that change in weather. Such information was of great importance to the Celtic public. Discussions of the events of the weather were chewed over at the kitchen table along with the daily bread. Nothing ever changes: today, too, a farmer's fate still lies in the skies.

Mothers came first in the Celtic world, and the aspen always hovered close to home turf, near the house or the ditches that edged the fields. Somewhere in the deep past the tree was given another name, a nickname, *Eabha*, or Eve, and was known as *crann eabhadh*, or Eve's tree. But Eve, as the mother of mankind, was also known as a complainer and nagger. Because the aspen made a fairly constant noise, it was called *crann creathach*, or the doddering tree. In the south of Ireland where the Celts were more highly strung in tune with the weather, the aspen was said to be a *cnámhseala*, or old woman, stirring about in her own bones. This was an insult a person might whisper about their mother-in-law.

The medicines of the aspen were well known to Druidic physicians. Fourteen or so forms of salicylic acid are produced by all parts of the tree. Many of these medicines have been lost, but one that has survived is still used by the beekeeper. When honeybees are angry because of humidity or if they can smell adrenaline from somebody's skin, they will sting. A mature aspen leaf can be crushed to release salicylic acid, which, when held for a few minutes with a little pressure on the skin as a green bandage, can relieve the pain of such a sting.

The oldest medicinal use of the aspen comes from the boreal north of Canada under the living pharmacopoeia of the Algonquin Cree and Ojibwe peoples. The aspen is considered to be an anti-famine tree. When trappers or hunters run into trouble in this enormous landscape, they can turn to the aspen for food. They peel a section of the bark off the main body of the tree to reveal the greenish cambial layer. This sweet food, which tastes of melon, is a survival meal.

The Celts put the grumbling aspen, their *Eabhadh*, into their alphabet as a *bile*, or sacred tree, and it became the Ogham script for the letter *E*. This letter was denoted by a vertical line crossed by four horizontal lines.

In recent years botanists have discovered an astonishing aspen in the United States whose roots survived the Pleistocene ice age by vegetative reproduction. The cells cloned themselves underground to form a root mass of almost two hundred acres, making the aspen 1.6 million years old, the oldest organism in the world.

I discovered something else about the aspen when I was up in the boreal forest. To take a break while filming, a few of us went fishing in the crystalline, potable waters up there. What I mean to say is that five of us were fishing for freshwater pickerel and I was just pretending. I kept my hook and bait within an inch of the surface and carefully watched my line to make sure it was out of reach of any fish. My antics did not go unnoticed, especially by Sophia Rabliauskas, the medicine woman at my side, who would occasionally let out a chirp of laughter.

Suddenly from one side of the boat a beaver, with only its nose showing, came swimming upstream with a massive branch of green-leafed aspen in tow. It jammed the aspen into place in its lodge and vanished. "That's our medicine," Sophia said. "Eat the beaver that has eaten the aspen and you get all the cold and flu prevention you need. This is bush medicine for our people."

F

Alder
Fearn

When I returned to the Lisheens Valley every summer, my mantle of schooling fell away as soon as I climbed off the bus and up beside Pat on the farm cart, by the clip-clop of the horse's hooves, the rushing water of the river, and the piquant smell of animals mixed with honeysuckle.

I felt like a queen sitting on my Macroom oatmeal bag, my young eyes sovereign over a landscape that felt like mine. The alders stood in silence as I passed, the two cartwheels bouncing me along up into my hills.

These wheels had changed everything for the ancient Celts, who took to this incredible invention and made it their own. The smiths, who ranked with the intellectuals of Celtic society, examined the wheel with the intention of improving it. They added a continuous band of hot iron to the outside of the wooden wheel. On cooling it contracted, which melded the iron with the wood. Then the smiths hammered in some nails to give the wheel more grip on the road's surface. This new vehicle, called the *carr* by the Celts, could now carry freight along with speed. That was the beginning of some familiar headaches—traffic, quickly joined by road maintenance, as always, pinned to the bullseye of its cost.

As traffic increased throughout the Celtic world, highways became necessary, then superhighways. Fords, bogs and wetlands were the challenge in designing these roads. The flood plain alder tree, which would not rot under water, was the answer. Its uniformity was perfect for building roads and highways. The community of these trees growing by the water produced a log of about 130 feet in length with a working diameter of three feet. Logs of thirteen feet made the road bed. Depending on the terrain, oak, elm, hazel and yew were deployed, too, their round upper surfaces flaked off by an adze. The new road could accommodate the comfortable passage of two wagons or two chariots—another evolution of the cart—side by side.

This network of highways went across the land for miles, but more importantly it crossed bog land where the road's condition was reinforced by fact and by law. The Brehon Laws protected these highways, along with the older *Seanchus Mór* codes of law, which stated that the king was responsible for the road maintenance of his territory. If travellers were to be harmed by a lack of upkeep, then the king was responsible to pay them or their family compensation for their injuries. However, if the traveller were to damage the highway due to his own carelessness, then he or his family had to pay just compensation to the king or chieftain.

When a highway needed to cross a river, the precise construction of a safe bridge, called *droichead*, was also dictated by the *Seanchus Mór*. Included also were rules for the construction of culverts.

There are the famous five highways, or *slite*, that still remain in Ireland from ancient times. Many of the important *slite* elsewhere in Europe became the beds of Roman road construction. The *Slí Asail* ran in a northwesterly direction into the Tara palace of the High King. The *Slí Mudluachra* cut through Tara, one section going north and the other south. Then *Slí Cualan* ran southeast through Dublin and *Slí Dála* southwest from Tara. The most famous *slí* was *Slí Mór*, a big highway that ran from Tara to Galway mostly on a sand esker. This was the favourite route of Princess Niamh, daughter of the last High King, whose red hair flashed to catch up with her love of speed. She is the model of prowess, power and pride for the women of Ireland.

The alder used in the ancient highways, *Alnus glutinosa*, is a tree native to Europe, North Africa and many parts of Asia. This member of the birch family, with its waterproof curtains of male catkins, is a special prize of spring, loaded with tannins that ooze like blood when the tree is cut. The air oxidizes the white wood to red in minutes and forms the basis of the superstition that it is unlucky to cut into this tree. But that reaction is the chemical basis of the alder being used as a dye, set using various vegetable mordants into yellow, reds and pinks, and black, green and brown.

The medicines of the birch family of which this alder is a member were not exclusive to the Druids, but known and used all over the civilized world. One of the earliest uses of the alder, *Alnus glutinosa*, was as an anti-microbial and analgesic mouthwash. The inner cambial layer of the bark was used, that is, the solid, damp tissue just inside the bark. A decoction, using some of this tissue, is used as a mouth rinse that is spat out after a few minutes' flush. This stops pain and inflammation of the gums and mouth. Interestingly, First Nations used a similar bark decoction of the different North American species to relieve the searing pain of scalds and burns.

A decoction of the bark mixed with mature green leaves of alder was used throughout the Celtic world for pain relief. This solution was applied as a wash on joints such as the knee and elbow and painful hands, and then allowed to air dry. People also drank a spring tonic tisane of the bark and immature leaves of the alder.

The Druids regarded the alder as a *bile*, or sacred tree. It was blessed with another name, *Fearn*, for the letter *F* in the Ogham alphabet. That *F* was denoted by a vertical line met by three horizontal parallel lines to the right. The ancient Druids also believed that *Fearn* was the guardian of water, a substance both sacred and holy.

G
Ivy
Gort

In ancient woodlands, ivy is the silent climber of vertical spaces, using the trunks of the trees to get to the sun. Called *gort* in the old tongue, this woody perennial begins life as a fat, shining black seed edged into the soil at the base of an oak tree. Time goes by, five hundred years and more, and the plant that springs from that seed reaches the canopy. The Celtic world regarded ivy, *Hedera helix*, as a magical plant that bestowed protection against evil spirits. Each household made use of its evergreen leaves when the sun dipped to the lowest place on the horizon in December, *Mí na Nollag*, hanging swags over the fireplace mantle to bring this protection into the house—a custom that has lasted in the west until today.

In late December, mummers, called straw boys or *cleamairí* for the ragged clothes they put on for the revels, used ivy as part of their disguise as they went from house to house play-acting, singing and reciting poetry for their neighbours. Like us, the Celts wanted a clean slate going into the new year; among the mummers, the men would dress like women and the women like men, and if they could fool their friends and relatives as to who they were, they would fool the spirits too. If one of them was correctly named, though, he or she was condemned to repeat the mistakes of the old year.

The Greeks dedicated the ivy to Bacchus, their god of wine, because ivy was a cure for their excesses. Even though the leaves were toxic, they were used as a preventative treatment for drunkenness, the young leaves infused into wine. This claim to faux sobriety was passed on to the English, who painted great swales of ivy over the doorways and the names of their favourite taverns.

You can still see great oak trees with girdles of five-hundred-year-old ivy shooting up into the canopy in the Raheen Wood in Ireland—a tiny piece of ancient woodland left over from the hunting days of the last *Ard-Rí*, or High King. It's the only place on the island that shines a light on the environment the Druids moved in and the kinds of plants that fed their medicinal endeavours. At a distance, the stems of the ivy climbing up these old trees look like muscles. Each tree is a powerhouse of growth, seemingly undiminished by the ivy that wraps around it. Perhaps it benefits the tree by supplying auxin, the growth hormone of the plant world, keeping the oak healthy.

As your eye stretches up the tree, following the ivy, you'll see that the shapes of the leaves change as the plant grows higher, clinging to the tree with a series of horizontal roots that act like claws. The leaves lowest to the ground have multiple lobes; as it climbs, the leaves lose those indentations. The youngest leaves, at the top of the plant, bear clusters of small greenish flowers in the late autumn that, once fertilized, produce black, globose fruits as toxic as the leaves.

Like many other poisonous plants, ivy is medicinally valuable. It is a member of the ginseng family whose enigmatic medicines are hard to understand because they act at the cellular level. The Druids used the unlobed leaves growing near the canopy of the oak in the treatment of various aches and pains, though the exact recipes have been lost.

The leaf was also used as a poultice for rheumatic pain, the black resin extract in dental work, and a leaf rinse acidified in vinegar as a mouth-wash to ease toothache.

This ginseng of the Celts had its place in the Ogham script even though it was a woody perennial, not a tree, and it was given the letter G for *gort*. That word had deep significance, meaning "field" and also, as *gorta*, "famine and hunger." Their lives depended on the fine balance between what could be cultivated in the field and the scourge of famine.

In the script, the letter G is written as a long vertical line intersected by two short parallel lines declining from left to right. Its guttural con-sonant flavours the Irish tongue in poetry and song. *Gort* was sacred to the Druids because it was medicine and a protector of their forests, a two-hundred-foot plant rising to the sun, moon, and stars. Temperate zone ancient woodlands are the only places to find this particular ivy, but the chainsaws continue to hum.

H
Hawthorn
Huath

The ancient Celts understood the hawthorn as a purveyor of power. They believed the tree was one entrance to the world of the *síoga*, the fairies, the people of good deeds, or *na daoine maithe*.

Druidic scholars knew the night skies and the solar system and described all of their peculiarities in mathematics. They developed the Coligny calendar. They worked with iron and practiced sustainable agriculture. They invented and reinvented the democracy of just laws they called the Brehon Laws. They put their oral culture into a written form called the Ogham script. And the Druids schooled both genders equally.

Their belief system was governed by the concept of the soul, *anam*, and the spiritual guidance, *anamchara*, that arose from everywhere. They believed that the living world was filled with soul, from the water to the mountains, to the grass, to the wild animals and insects. All of life was connected by this soul and, because of that, life in all of its forms needed to be protected.

Anam spread into the afterlife, like a great sheet of consciousness. The fairy people lived in this parallel world and could come and go at will. They announced to the living the upcoming deaths of all members of the ancient Celtic families whose names began with Mac or Ó. These royal families received a visit of the *beansí*, or banshee, a fairy woman, who could produce *ceolsí*, enchanting music, or *solassí*, fairy lights, as a warning of death.

The hawthorn is a member of the rose, or *Rosaceae*, family. Its name in Latin is *Crataegus monogyna*. This sparse, spiny tree can grow up to thirty feet. The wood is a pinkish colour and is incredibly strong. In May, the tree bursts forth with a halo of bitter-smelling white flowers, bunched in clusters. They are followed by ovoid scarlet fruits that are technically little apples or pomes, usually called haws, which become sweet after a frost. They were the trail foods of the Celtic farming community in the fall, exceptionally good for heart health.

The magic of the hawthorn is an ancient medicine, which is still in use in today's operating rooms. Its commercial extracts go by the trademarked names of Curtacrat, Crataegus-Krussler and Esbericard. These medicines are cardiotonics that act as vasodilators. They work on an unusual target area in the human heart—the left ascending coronary artery, which is the transport system for feeding the muscles of the heart. The hawthorn extract opens that coronary artery, allowing a greater blood flow to oxygenate this vital muscular pump.

Druidic physicians in ancient times made good use of hawthorn medicines for unspecified weakness. Today the extract is prescribed to treat hypertension associated with myocardial weakness, arteriosclerosis, tachycardia, and some of the problems associated with angina pectoris.

The mature leaves of the hawthorn carry another potent magic, a growth hormone for the butterfly world—a compound that pumps power into the caterpillars as they feed. This, in turn, strengthens the butterfly population, which aids in their ability to pollinate and cross-pollinate plants in the wild.

Observation was the key with the Druidic physicians. What they could not explain they called magic. Their observations have stood the test of time. They stand with us, now, as cornerstones of biochemistry.

The hawthorn was given a name for its sacred status, *huath*, which became the letter *H* in the Ogham script. *H* is designated by a vertical line met by a single horizontal line to the left.

A couple of years ago, an extraordinary thing happened to me in an international airport. I was sitting quietly waiting for my flight, when I noticed an agitated Chinese woman standing near me. She had her boarding card in one hand and her passport in the other. I heard her flight being called and realized that she did not speak English and was beginning to panic. So I walked over to her and with gestures I managed to direct her towards the gate on her boarding card. She uttered words of thanks in Mandarin and started to root around in her baggage, unearthing five pink sticks that she gave to me in gratitude. Later that day, I asked a friend to translate the Chinese lettering on the sticks for me. It turned out they were hawthorn, a different species from the European hawthorn. The Chinese species was a travel medicine for vasocontrol.

At the time, I was working on research on hemodilution and hawthorn myself. Was that a coincidence?

I
Yew
Iúr

Cut down a forest. Cut all the forests down—and you destroy the spiritual life of a woodland culture. This is called genocide, the systematic destruction of a cultural group, blow by blow.

The Irish yew, *Taxus baccata*, the ancient tree of bereavement of the Celtic culture, was extirpated from the lush lands of Ireland by the English. The forested landscape described by the courts of the High King of Tara as being *iúrach*, or abounding in evergreen yew trees, was razed to the ground. The tough, close-grained, bendable, watertight wood was used to craft weapons of war and by the woman of the house, *bean an tí*, in her dairy.

Long ago, the rose-red wood of the yew was a widespread industry for the Celts. The close grain, with its tiny pores, was made waterproof by milk and its fatty products. Under damp conditions, the wood did not decay and remained wholesome. Yew was used in the churns, the milk vats and the wooden treenware for making butter. These yew vessels were called *iúróirí* and the specialized carpenters who made them were called *iúróirí*. The entire specialty was *iúróireacht*, or yew-work. The yew wood, over the years, absorbed the patina of the cream the way oak barrels age whisky and specialty artisanal butters were swapped and sold by women far and wide.

The wood from the mature yews, *Taxus baccata*, when grown in rich soil, has exceptional qualities. Since the internal plumbing of tracheids is flexible and retains great strength even when bent in shape, yew wood was said to be the best for making bows, the principal weapons of ancient warfare. The Celts were known for their hand-and-eye coordination. They were the elite of the military, even in Rome.

By luck or by chance the yew returned to Ireland in 1780, still during the penal times. One spring morning near Lough Erne in what is now northern Ireland, a farmer was doing the rounds. His eye caught something strange growing in one of his fields. On closer inspection he found two small evergreen yew saplings growing stoutly side by side. His land had long been cleared of forest for the plantation English, some of them Beresfords, to take over. These two yews, almost miraculously sprung from seed that had lain dormant, were all that remained of the yew forests of ancient Ireland.

The yew, *Taxus baccata*, is a medicinal tree, as are the other seven species in its genus (though some botanists consider all eight species to be just one). Across the globe this renowned species is a tree of bereavement to almost all cultures, its wood used in coffins, its branches twined in funeral wreaths. The tree is poisonous, with one strange exception. The bright-red fruit, called an aril, is composed of a bony seed that sits inside the pulpy flesh. The pulpy red flesh is a choice bird food that seems to be non-toxic.

Otherwise the Irish yew and all the yews across the world are poisonous to cattle and to people. But they produce an extraordinary family of biochemicals called the taxols, which are currently used in the treatment of many cancers. Like many medicines in this arena, a small dose is a cure, while a large dose is a deadly poison.

The Druidic physicians were aware of the yew cures and placed the tree firmly in the land of the living as a *bile*, or sacred tree. In Celtic households, butter stored in yew containers was put on burns and scalds to exclude the air from the burnt surface until healing began. Milk stored in yew vats was boiled with onions as a medicinal drink against colds and used as a body sweat against viruses. Buttermilk from yew jugs was drunk in the spring as a tonic and also to clarify the quality of the skin of teenagers, especially those afflicted with minor eruptions, redness and scarring, and was sometimes applied around the skin of the eyes to relieve facial tension.

Many of the more valuable medicines of the Druidic physicians, using bark, root bark, aril, wood and foliage, have been lost. These secret medicines were given as part of an honour system to important families to hold into the future. They went underground for five hundred years during the Penal Laws only to re-emerge in the late nineteen hundreds. Then they were thrown out as useless folk medicine in favour of modern pills. A number of these medicines were cures for cancer. Many were used in pain management, as well.

But some old cures remain. The Iroquois peoples of North America used yew extract from their species, *Taxus brevifolia*, as a synergist for their medicines. A synergist is a biochemical boosting mechanism that makes a medicine more effective. Yew was traditionally used as a steam bath treatment for pain. The foliage in water was taken to a boil, and the steam was then used to induce perspiration. The taxol extract landed on the skin, dried there, and relieved the chronic pain in joints. The yew was also used to reduce numbness of limbs and to reclock menstruation.

The Druids, in their wisdom, put the letter *I* for *iúr* into the Ogham script. The letter is a vertical line intersected by five horizontal, parallel lines. *Iúr* was their word for yew in this alphabet of trees.

The story of the yew has an ending in my arboretum with a species that I thought had been lost in this area. It is the ground yew, *Taxus canadensis*. This plant is a leftover from the ancient North American forests of millions of years ago. I found it crouching in the shade of a large tree under a cedar rail fence trying desperately to grow. And it will, it will. While it lives, there is hope it will yield more medicines for cancer. That is biodiversity in action.

Ng
Rush
Brobh

Lighting for the Celts came in the form of a plant. It was not powered by a nuclear reactor whose spent rods will remain radioactive as long as the half-life remains, ticking the millennia away. It was a rush, a plant of the bogs, a wild creature of the wetlands.

Every Celt, even the children, knew how to make a rush candle, *coinneal feaga*. The process was simple and entirely sustainable. The genius of the rush lay deep in its own history as a life form. At a time when dinosaurs were filling their bellies, the rush, then a monstrous plant, was on the menu. It needed to make some internal changes to survive the hot, wet climate. A sponge became the central living core of the leaf, an organ tissue called the parenchyma. The parenchyma holds air spaces filled with oxygen, invisible to the eye but essential to life.

Celtic households sent their children out as scouts to find the best plants. Usually, the tallest hovered at the edge of good pasture land and a wet field. The growth pattern of the rush is circular, the younger leaves growing on the periphery. The longest leaves, found in the centre, can often have brown whiskers, which are the sexual organs of the plant. The leaf is waxed and waterproof. The shape is that of a very elongated green cone. A quick, sharp pull at the base of the plant harvests each leaf, which should have a white, bleached region at that base. Care should be taken in harvesting not to bend or injure the leaf.

Then the real fun begins. You place the leaf on a flat board and cut into its green surface from top to bottom with a sharp knife, going only as deep as the cuticle. Then beginning at the white base, you use your thumbnail to peel the snow-white parenchyma away. This parenchyma, with its air pockets, becomes an ideal non-drip candlewick. You dip the wick into hot fat (mutton fat was used in the past, as it stores easily in a solid form). When the wick cools down, you dip it again and again. A well-made candle could burn for about one hour. The candles or tapers were placed into special holders, one at a time, to burn. They produced a good quality light indoors, adding a lovely soft glow to the evening's chores.

The common soft rush of the Celts, *Juncus effusus*, is one of several hundred species of the rush family worldwide. It was also used to line the open-pit storage of potatoes, with other vegetables such as cabbages on top, for the winter months. It was used as underbedding for milking cows and poultry during wet weather as a form of waterproof drainage. Rushes were woven into beautiful floor mats, bedding, chair seats and other household decorations. They are still in use today as the fragrant tatami floor coverings of Japan.

It has long been known in the farming community that the rush is poisonous to cattle. When grass or hay is available, cattle will usually graze around these plants, ignoring them completely. But when hunger descends, the animals forget their previous indifference and begin to gorge themselves on rushes, unable to stop eating them until blindness, coma and death overtake them.

The poisons in the rush are not understood, but they were used as part of an athlete's cleansing prior to a run or to playing many of the ball games of the ancient world. Athletes would wash themselves three times with rush water to gain endurance for the stressful task ahead. This body wash was thought to be some kind of skin emetic.

The rush may well have aided the greatest long-distance runners of Ireland. Fionn, the fair one, and his band of warrior poets, the *filíochta*, ran the length of Ireland in a single run, accompanied by a faithful pack of Irish wolfhounds braying at their heels. Such feats of endurance are the weft of many Irish legends.

And so the rush was placed with the trees as having a sacred status and given the nasal-sounding letter *ng* in the Ogham script. The letter is known as *ngetal* today. It is seen in writing on the Ogham stones as a single vertical line intersected by three parallel lines declining to the right. But the rushes' greatest treasure was the light they shed on the ordinary life of so many Celtic homes.

L
Rowan
Luis

The Celts are not the only people who looked at the rowan tree with a gulp of fear. In many parts of the global forest, folklore and superstition abound surrounding this tree, even in modern times. The rowan was used for exorcism to release spirits, the anima of life. Amulets of the wood or sometimes branch tips were worn as a protection against any stray evil that may come wandering down the road.

The Celts, however, had another twist on these legendary powers. They considered the rowan to be a tree of enchantment. It was believed that the fairies were deeply enamoured with the beauty of the rowan's snow-white, May-blooming flowers followed by bright-red berries in the fall. It was said that these creatures intoxicated themselves with the brick-red juice of those berries. During the dark hours of the night when sleep cast its spell on the human family, the drunken fairies got up to tricks—and some of them were not very pleasant. The Celts, in turn, had a saying for their enemies: *"Caor thine ort!"* or "May the fire of the rowan berries burn you." These words did not spring from the heart of charity.

The farming community of the Celtic world used the rowans in their hedgerows as a sort of solar newspaper. The date the blooms broke out was carefully noted and applied as a measure for the expected grain harvest in the fields. The quality and quantity of the rowan berries was viewed as an indicator for how large the grain kernels might be. If the rowan fruit matured to a proper rosy colour, all was well in their world.

The rowan's ability to enchant was used by Druidic physicians to enhance the lulling effect, *sámhnas*, they sought to achieve with the recitation of verse or reading from holy books. To be in a state of *sámhnas* meant that you dropped your guard of vigilance, as an infant does on hearing a lullaby. This happens also with animals. Many songs were used in the past to relax milking cows to increase milk production. The Druidic physicians considered that this lulling gave the mind a rest, which accelerated a person into a greater state of calm.

As with meditation, in a state of *sámhnas* time changes and a minute can become ten—or the other way around. Such a holiday of the mind is beneficial for the relaxation of the entire body, and we now know that this is true particularly for the adrenal cortex. The Druidic physicians believed that being mindful was a health benefit. The rowan was a *sámhnas* medicine.

The Celtic rowan, *Sorbus aucuparia*, contains stimulants, too, as do the other eighty-five or so rowan species worldwide. The effects of the rowan stimulants are on a scale similar to that of coffee, tea and other such plants. The ripe, raw fruit of the rowan is poisonous to the human family, so, like the elderberry, it must be cooked before eating. A tisane of the rowan was drunk by the Celts as a tonic. The exact ingredients of the tisane, as well as records of the species and maturity of the rowan used by the Celts, have been lost. The closest rowan tisane that is still being drunk is made by the northern First Nations peoples of North America, but their rowan, *Sorbus americana*, has much stronger medicine. An infusion of the inner bark of a rowan, harvested in mid-spring, along with the sweet flag, *Acorus calamus*, is their usual tonic.

The last peoples to use the rowan as an important medicine in their traditional pharmacopoeia are the Slave, Cree and Chipewyan. They use three native rowan species of the boreal, *S. americana, S. decora* and *S. scopulina.* The trees in this northern region are reduced to tall shrubs, and their medicine is made more potent by the stress of short days combined with chilling temperatures.

These three nations call the rowan the "medicine stick." The upper green pinnate leaves, when mature, are used to make a decoction to treat colds, coughs and headaches. A root decoction is used to treat lumbar back pain. In a very complex combination with other native herbs and plants, the rowan is also used in the treatment of diabetes and cancers.

There is little doubt that the Druidic physicians dedicated the Celtic rowan as a sacred tree because of its medicinal value to their society. The rowan was called *luis* and given the letter *L* in the Ogham alphabet. This letter was designated as a vertical line met by two horizontal lines to the right.

M
Blackberry
Muin

On his way to get the cows of an evening in early autumn, an Irish farmer will plunge a rough hand into a mat of thorny canes growing in his ditch and emerge with a clutch of blackberries. He'll only eat the berries with a snow-white base and toss the other ones away.

This ritual of autumnal picking of blackberries has been going on for thousands of years in the Celtic world. The species picked is the bramble or blackberry, *Rubus laciniatus*, which still grows wild in Ireland and Europe. It is a member of the rose, or *Rosaceae*, family, with a thousand siblings across the world, from the chilly climate of the Arctic to the heat of the Indian subcontinent.

The plant with its arching canes begins life as a dry seed that has travelled through the bacterial biota of the intestine of bird, human or mammal. The little satisfaction of nitrogen left over sets the seed up for life. The blackberry is a biennial. During the first year, canes produce food for the second year's crop. The little white rose-like flower gets fertilized and becomes green, then red, then black. The fruit is called a drupe by botanists, a series of seeds set in their own packages of water-tight skin that hold the juice—a temptation to both man and beast.

From a blackberry's point of view, survival is important. Like all other species, it is involved in the dance of reproduction. The choices are limited for a plant that wears barbed-wire stockings and whips the air with thorns that rip open human flesh. But the sweet fruit is a temptation that causes the picker to risk their flesh, and once you pick and eat the fruit, health is on its way.

Biochemists today have pinpointed a biochemical in the blackberry called ellagic acid. It is a phytochemical regulator, an immune system booster, that appears to offer protection from some forms of cancer, which is under investigation in Canada. It has long been suspected that the juice has anti-diabetic activity, too, but that has not yet been proven through clinical trials. A phytochemical regulator is the cellular guidance system in a plant that sometimes crosses over into food and acts in the human gut. The plant hormone influences human pathways, too, including the farmer with a sweet tooth searching for berries as he brings the cows home for milking.

One of the jobs on the farm in Lisheens during my Brehon days was to bring in the cows. I was a famous spoiler of animals and always had an audience for my task. Around four o'clock, I picked up the walking stick. The cats and kittens took the cue and soon the horses noticed me, too, and came at a canter. The dogs were always glued to my side, and now they ranged ahead full of purpose. The laying hens also followed, at least for as far as they felt safe. The sheep and donkeys joined in, along with a few turkeys, and we all set off to fetch the cows down in the valley. Strawberry, my favourite matron among the cattle, always came first to greet me. Then we all proceeded back to the farmstead, stopping every now and then to pick and eat blackberries. I dearly love the fruit, and back then my black fingers, tongue and lips were ample proof of my malingering.

The medicine men and women of ancient and modern Indigenous cultures would call what I was eating "bush food." They understand that when their peoples stop eating such bush food, they lose their health. Wild food of all kinds, from untainted, pure, genetic sources, have a phytochemical regulation system that modern science is just now trying to understand. Failing to pay attention to how important these foods are to health has led us to an epidemic of obesity, intestinal troubles and diabetes. Somehow the Druidic physicians understood this by simple observation, which is, after all, the foremost tool in problem solving.

The Druidic physicians were well aware of the scarcity of medicinal species even in their own time. They had heard from the Greek physicians around the time of the birth of Christ that a form of giant fennel was going extinct. This fennel, called *silphium*, grew on dry mountainsides in a specific banded layer of soil near the coast of Libya. Used as a contraceptive, this plant was in much demand across the world and had become more valuable than pure silver. By AD 37, as the story is told, only a single stalk was left.

Physicians from ancient cultures laid down the groundwork of preventative medicine a long time ago. Knowledge was power, then and now. Medicine was passed around from culture to culture in a system of barter. The evidence lingers in the language. My family name is Beresford, which is *Dún Sméarach* in Irish. We are the people of the "fort of the blackberries"—the one plant that refuses to grow in my garden!

The blackberry is a species of the wildwoods. It likes to grow on the sunny edges of the forest. Natural wood decay feeds the canes with the potassium and potash necessary for the growth of fruit. The songbirds nesting in the trees come out into the open spaces to eat and then return to the edges of the forest to sunbathe. When they relax, their intestinal muscles do, too, passing seeds. The Druids understood this ecosystem and protected it by granting the blackberry a sacred status.

The blackberry plant, called *muin*, became the Ogham script's letter *M*—represented by a long, vertical line crossed by a single line tilted up to the left.

N
Ash
Nion

To the ancient world, the ash was a mystical tree. Many cultures regarded the ash as the primordial being—the tree that was first created from nothing.

The Wabanaki peoples speak of Glooscap, their creator and great warrior. Glooscap made arrow shafts from the heartwood of the ash, running them through a stone borehole in the age-old tradition. Then he hardened and toughened these arrows, picked out one and shot it at a mighty white ash, *Fraxinus americana*. The arrow made a hole in the heart of the tree. The first members of the human family poured out of this hole in a living stream to take their place on the planet.

In a Norse saga, the primordial tree was also an ash, called *Yggdrasill*. This heroic tree created the souls of the unborn and protected those souls in its mighty limbs. The ash cared for them as long as they were embryonic and, when they were ready, released them into the world.

The Hindu faithful consider trees to be great luminaries that have reached the highest plane of the plant world. In many parts of India, the faithful adorn trees with colour and dedicate rice offerings to them.

The Celts recognized the divinity of their ash, *Fraxinus excelsior*, as it towered above the crown canopies of the oaks. The dark, ribbed trunks headed for the heavens. The tree with its rattle of dry samara seeds was a magnet for foraging songbirds. The nightingale unfolded musical verse from its limbs. The wood was like no other in the drizzle of that climate. It could hold a fire even if cut still green because its fibres were laden with oil from the heartwood to the twigs. Ash and peat created a natural marriage of warmth in the Celtic firepits.

The Celts played a unique game, called hurling, using ash sticks. I come from a family of hurlers, some of whom were famous. My uncle Patrick, who was called Rocky Donoghue in his hurling days, had a huge following and ran like a rocket. But I remember him telling me something very interesting about the game and about sports in general. He said that the greatest danger to democracy was disenfranchised young men on street corners; they are a tinderbox for conflict. The ancient Celts used sports to give the youth direction and a test of their mettle, to tamp down sentiments towards war.

It is not surprising that the master poets, learned men and expert thinkers of the Celts gave the ash its due. They honoured the ash for its noble stature in the landscape, and for the way it bound their culture from top to bottom through the fierce and fast game of hurling. Hurling was a game of honour with nothing whatsoever to do with material wealth, in which athletes were singled out for their humility in winning.

The sacred ash was named *nion* in the Ogham script and given the letter *N*. This letter was written as a vertical line met on the right by five parallel horizontal lines.

The ash may have been a source of medicines used by Druidic physicians. If so, they have been lost over time. Both the Celtic ash and the North American ash carry a similar biochemical called escin. This chemical tightens the peripheral arteries of the skin. The First Nations used their ash as a preparation for hunting. Before setting out, the hunter washed his body down with a decoction made from the mature bark of the ash, which blocked his smell from rising from his skin so that he would be undetectable by his prey, even when upwind of it.

It is also very possible that the manna described in the Bible—the food supplied to the Israelites as they walked through the wilderness on their way to the Holy Land—was also from an ash. In that hot climate the flowering ash, *Fraxinus ornus*, grows. This little fragrant tree is like the Canadian maple in that it, too, has an excess of sugars. Its sugars are endurance sugars, ideal if not perfect for people on the run. The sugars can be extracted by puncturing the bark of the ash in a long vertical slash. The sugar begins to bleed out immediately, accumulating as it dries into a white mass at the bottom of the cut. This "manna" contains four complex sugars and a number of medicines. This tree is still grown for the purpose of producing a food called manna in Calabria in southern Italy, where the tree is considered to be a crop tree, like its cousin the olive.

O
Gorse
Aiteann

There is a feast of the fields in most of Ireland, though in this era of tourism it seems that only the visitors notice the yellow of the gorse in bloom. The dwarf gorse, *Ulex minor*, supplies the main feast, flowering from June into December with a September peak. The tall gorse plant, a moping shrub, is *Ulex europaeus*, the European gorse. They are both of the pea family, and their flowers are an acidic yellow. Yellow, especially if it fluoresces in early morning and evening, is an invitation to the pollinators on whose efforts our daily food relies.

The yellow flower of the gorse looks as innocent as a lightbulb, but its arrangement is complex. It is a snapdragon shape. The tall plant bears big flowers and the dwarf accommodates smaller ones, the perfect size for a honeybee. The tall variety flowers earlier than the dwarf version, and so the bees sneak off a load of pollen while the dwarf below is still asleep and not producing nectar. This is exactly what the bee needs to bring back to the hive, pollen that is itself a tonic mixture for the growing female bees in the wax chambers.

In the distant past, the gorse was the friend of the Celtic farmer, also. The two shrubs required almost nothing by way of soil. They thrived on a strict starvation diet of sharp alkaline sand together with a splash of rain. They fitted nicely into the hedgerows and commonage as dividers for fields. In the company of dry-stone walls, the gorse became a sounding board for the seasons. Early flowers meant the fields could be worked early. If that early flowering was followed by a long and continued blooming period, the farmer could expect a good harvest, since the hives would be filled to capacity and could easily pollinate all of the crops.

The ancient Celts drew a botanical line down their species of gorse. Maybe this had something to do with their agricultural practice of fallow rotation with nitrogen-rich crops. Or maybe each growing region had to be treated differently. The dwarf gorse that was native to Ireland was called *aiteann gaelach*, while the large gorse was called *aiteann gallda*, or the foreign gorse. While both species grow in Ireland, it is the dwarfed plant that was so sacred to the Druids.

The Druidic physicians used gorse for their medicines. Its prescriptive formulas have been lost, but gorse honey, still considered a healing honey, contains an immense diversity of biochemistry. Dwarf plant species live a different chemical life to the larger kinds. They are regulated by a series of plant hormones called gibberellins. One of these is an outstanding growth regulator called gibberellic acid, based on a chemical nucleus that is almost identical to the hormone regulators in humans. Plants with a dwarfed genome hold a treasury of more effective medicines than those of their lanky brothers and sisters.

If dwarf gorse was rare, it would be a collector's item in the botanical world. Probably if it were not so abundant, it would have long since been noticed by at least one drug company. Over the millennia it has learned to survive soil poverty with the aid of specialized bacteria, *Rhizopus phaseoli*. These bacteria form external nodules on the root hairs that cleverly fix nitrogen, collecting it to feed the plant. This give and take produces the miracle of lectins, molecules that have enhanced anti-tumour effects and may reduce rejection and induce immune tolerance during organ and tissue transplants.

Indeed, the Druidic physicians were skilled surgeons two thousand years ago. This fact was known and recorded by the Roman Empire. Druids ran specialized hospitals in many parts of the Celtic world; the word "hospital" itself is one of ancient Irish descent. They used gorse field honey on open wounds on the battlefield as a natural anti-bacterial for the skin.

Everyday Celts collected dead plants, or furze, to dry and store as kindling for turf fires. Dead furze has a low flashpoint and will blaze in minutes. Furze was used as a mobile field boundary throughout the ancient world. The plant also provides grazing for donkeys and horses.

The Druids called gorse *aiteann* in the new Celtic alphabet, giving it the letter *O* in their Ogham script. It was written as a vertical line intersected by two horizontal parallel lines.

As the common becomes uncommon, maybe the true medicinal value of this legume will be uncovered.

Q
Apple
Úll

The apple was part of the ancient wildwood of Ireland. The fruit was green and bitter—a large, ugly crabapple, a little bigger than an egg. The small trees sat at the edges and in the clearing of the oakwood forests. People and animals ate the fruit in the fall. The apple seeds found an opportunity for regeneration in the damp, rich ground.

There is nothing easier to plant than an apple. The slippery, dark seed is designed to slide easily out of its loculus embryo at the core of the apple. The pointier tip, a bit smaller than the other end, enables this birthing passage of the seed. The belly of the seed hits the ground, where its cyanide shields the seed from being eaten. Winter comes and blesses the seed's embryo with dormancy. The spring sun warms the dark seed coat, the testa, and an internal clock of chemicals wakes up and starts to tick down. The small apple pip puts down a snow-white radicle, or root. Then the plumule, the juvenile shoot, heads for the sun, still wearing the black cap of the seed coat for protection. As this baby stem elongates from under the cap, the two thick, embryonic leaves open to embrace the air, flinging the cap away. An apple tree is born in a brief sixteen days.

Apples were important to the Druids because the apple blossoms fed the honeybees. Quite often honey was the medium for drug delivery, something to make the medicine go down. Comb honey was itself considered to be a health food; both the honey and a smaller amount of the comb were consumed. Often honey was stored in jars or boxes in the anaerobic conditions of bogs; it has emerged millennia later as the perfect food.

In the spring, after the first flush of protein-rich pollen, honeybees have a need for nectar and then resins. Worker bees require the pollen to build up the brood by feeding the laying queen bee. Then once the embryos begin to pupate into young worker bees, they need high-energy liquid honey to give them endurance for flight and for foraging. Apple blossoms perfectly fill this spring gap of feeding. The supply of food must come at exactly the right time or the hive will weaken from hunger.

The apple tree makes a chemical call to the beehive. Worker scouts investigate the source of the scent and taste it for its chemicals. Then they perform a dance of orientation at the hive to provide directions for the other workers to locate the apple tree. The blossoms open. Each bud reveals five white and pink petals that stretch wide to reveal their nectar. The hive is fed with a golden honey and the bees now have the brawn to cross-pollinate field crops for the entire growing season.

The apple was just part of the chain of reactions that happened in the ancient wildwoods. The young worker bees, reared on the apple blossom nectar, became strong enough to travel further afield. They flew into the oak woods where they found the polymeric resins that stitched the tender tips of the oak trees with protective glue. Most pollinating bees need this glue to wallpaper their hives; they use their mandibles to tear off the resinous glue and ball it for flight. This is very hard work for a honeybee. Sometimes they fail, dropping the dark resin on the landing board of the hive before they can deliver it. It is only by communal effort that a body of bees survives.

The Celtic Druids honoured the tapestry of life around the honeybee. These workers were considered to be an extended part of the family. Births, marriages, deaths and anniversaries were announced to the bees. Grief was always shared with the bees in a form of non-verbal communication. The scrawny apple tree of the wildwood was honoured as a vital sacred tree, a *bile*, to the rural lives of the ordinary Celts.

The apple tree, *Malus pumila*, held the sacred apple, *úll*. In the Ogham script the *úll*, standing for the letter Q, was designated as a vertical line met by five equally spaced horizontal lines to the left.

There are twenty-five species of wild apple, including crabapples, across the northern hemisphere, from North America to Europe to western Asia. All of them are now rare and some are endangered. They are members of the rose, or *Rosaceae*, family and they are extremely important to the feeding and consequently the health of insect pollinators. The world's food supply is produced in large part on the labour of these insects and all the other pollinators that were once protected by Brehon Laws.

The Druidic physicians used the apple for many medicinal purposes. These have been lost in recent years, forgotten by the descendants of the old Celtic families who for generations had held the prescriptions and dispensed them for free. All of these practices were conducted in secret under the radar of the Penal Laws during the occupation of Ireland, when even the ownership of some seeds was forbidden.

However, the Druidic physicians also considered the apple to be a commonplace health food, something a person should partake of, along with eating fish, forest and health bathing, and accessing the tonic effects of the sea and maritime algae at certain times of the year. These practices were thought to correct the balance back into health. The apple does have an invisible medicine found in the skin—the wax-like water-proof coating of the apple that allows it to grow and expand on the tree until the scissor hormone of abscisic acid cuts the apple stem in the fall.

When we eat the skin, an agent in it dissolves in our stomach acid and moves along into the bacterial habitat of our intestines, where it acts as an important emulsifying agent. This agent allows the metropolis of bacteria to do its work, breaking down food and fuelling the bacteria to feed the gut itself. It now appears that the health of the gut is in turn vital to the health of the individual—thanks to the humble skin of the fruit.

Apples have fascinated me since I was a small child. I once was playing hide and seek with two playmates, Mary and Marjorie, in their attic. I made it out onto a window ledge—forty feet up from a stone courtyard below me—and saw that there was a roof not too far below. I lowered myself carefully down to that roof to hide and discovered a large tin can sitting there. It was filled with soil. In the centre a small twig grew, which had no leaves, only buds. It looked so lonely to me, sitting all alone on the grey stretch of cold slates. Later I asked my friends' mother what kind of plant was in the tin. She told me that it was a little apple tree. She had taken a seed from a tree that grew on her family farm in the Irish countryside. Far from being lonely, the little tree was the object of much attention from a woman who wanted to propagate a reminder of her past.

R
Elder
Ruis

The Romans believed that Celtic women, especially the ones with such unlikely red hair and flashing green eyes, simply did not know how to behave. They did not purr like the loyal wife, lover or daughter. They were a disgrace filled with learning. They even led their men into battle. It was rumoured that they could deliver a baby in a snap. And many of them were beautiful, seriously so, which the Roman invaders could not help but notice.

One source of their beauty was a small tree—actually a series of tall bushes—that grew in the rich, wet soil at the edges of rivers and streams and were connected by underground suckers. In late spring or sometimes in early summer, a panicle of white flowers appeared, each like a flat pancake balanced on a pin. Then in no time at all, these flower shapes changed into heavy masses of purple-black berries. The weight of the ripe fruit bent the branches almost to the ground. The ancient Celts named the tree for its heavy weight of fruit as *trom*, or the tree that bears weight.

This little tree is the elder, *Sambucus nigra*, and the black fruit it bears is the elderberry. It has around twenty relatives across the world, some found in the semi-tropics. They are all poisonous, some much more than others. The elder has been used as a cosmetic for thousands of years. The women of Egypt applied it to beautify themselves.

One secret of Celtic beauty lay in the fragrant, delicate white flowers of the elderberry. The tree carries a remarkable oil, mucilage and some complex resins that blend into a facial wash that has been a beauty treatment since time immemorial. It strengthens and protects the fine capillary network just under the skin, which is then able to tend to the business of circulation. Crow's feet can be reduced with this facial wash, so that the skin becomes smooth and youthful again.

Then there is the secret to the flashing eyes. The ancient Celts had to work hard for their daily bread. In winter when the days were short, every last beam of sunshine was valuable, so they could finish up tasks before darkness descended. Elderberries cooked into a porridge or strained for juice carry a complex sugar called sambucin that helps the eye to adjust to darkness, which gave those who consumed the berries an edge against the dark. Elderberry is still in use today in the treatment of night blindness.

The trick with the elderberry is that all parts of the plant, with the exception of the fresh or dried flowers and the cooked black fruit, are poisonous. The fresh or dried flowers, fruit, roots and leaves, including the bark, were used for medicine, along with the branches' pith. In the past, the white elder flowers were added to a peppermint tisane, the drink of choice in the treatment of the common cold. The Celtic prescriptions using the various parts of the elder mixed with other herbs have been lost. But they can be reconstructed using modern biochemistry and the wild forms of the tree.

Innovation sparked by past wisdoms may well be crucial to our future, especially in the age of superbugs. Newborn babies of the

Iroquois nation were once sponged with a wash of elderberry flowers, except for their eyes, ears and nose. They were rinsed with newly expressed mother's milk, which is sterile by nature and has a buffering effect on newborn skin.

Then there is the story of the elder tree as a magical species. Search out the magic and you will almost always find medicine. People of the past believed that spirits lived within the folds of this small tree. The elder was never burned in bonfires, because the flames would scorch the souls of the spirits inhabiting the tree.

The black elderberry of the Celtic world was well placed in the Ogham script of sacred trees. It was given the letter *R* and was called *ruis*. On the Ogham stones the letter *R* can be seen as a vertical line crossed by five parallel lines declining from left to right.

As a ward of the Brehon culture, I was accorded the Celtic female cachet, which was considerable, and told to both love and respect the person I was becoming. By sewing acts of kindness and courage, I would reap the character I wanted to become and my heart would run a true course for all of my life.

The elders versed in the mores of the Celtic culture instructed me in the meaning of a special word, *buíochas*. It means, as best as I can render it, tender gratitude. The *buíochas* should be very high in each person, like a glass that is full. *Buíochas* is also a self-protection. You should carry gratitude in your heart for everything inside and outside your life and all the small things that impinge on your consciousness. The feeling of *buíochas* is like a medicine of the mind that holds your life together. The old saying *buíochas le Dia*, thanks be to God, reminded the human family that life itself is the greatest gift and should therefore be treasured in yourself and in all others.

S
Willow
Sail

There was one bed that was most important to the Celts. It was not the marriage bed, because that could be remade by divorce, by a nod of the wife, according to the traditional laws. No, it was the willow bed, *an saileán,* a bed made of herbs, bare in the winter, bright in the spring and used in the early summer.

Every household of the Celts made use of these beds. The willow represented a universal building material in the form of its bendable branches, called sally rods—weaving material for strong, multi-purpose baskets and panniers, depending on the size of the rod. The winter's turf was carried in willow donkey panniers from the bog. Finer sally rods were used as floor and household brushes.

The willow also provided an important pain medicine, and it produced a rose-tan dye for raw linen and wool. In a pinch it provided animal bedding and fencing for cattle and heifers. Roughly woven, it protected vegetables.

Poultry benefited from the willow, too. Fine green sally rods, called osier sticks, were woven into a wide laying basket that was finished with a very firm rim for the laying hen to use like a ladder. These baskets were used both for laying and broody hens in a henhouse or stable. The open weave allowed the hen's feathers to breathe in a warm enclosed space while the hen was occupied. It reduced feather mites, while increasing the hen's feelings of protection during a portion of the day when she was quite vulnerable.

I remember watching my grand-uncle Denny weave sally baskets for the house. His pipe seemed to move with his hands and to his a cappella hum, which sometimes broke loose into an ancient Irish song. He would say to me, "*Cailín*, watch closely how I do this." And I would move closer, the smell of his Virginia plug tobacco in my nose. Under his hands, the basket seemed to rise like magic with each rod added and twisted around the upright staves. Then he wove the top into place, bending and fitting the larger rods with both hands close together until he was satisfied. The wonder of something so beautiful made from almost nothing, from a simple sally or willow bush, was like magic. The sharp, acidic smell of the bark as he twisted it by hand is still with me. He always held the finished basket on high, like a trophy, waving it to show the world something new. Then he rummaged in his waistcoat pocket for his plug. He was the last of the ancient lineage of bonesetters. Neither of us realized at the time that traditions were so fragile.

The willow was part of the natural landscape in the ancient world. The wild plants like the willow were harvested when needed. The harvesting never went beyond the point of sustainability. The ancient rule for taking from the commonage of the tribal garden, like the practice among the Indigenous peoples of North America, was that there should always be sufficient left over for the seventh generation into the future.

The Druidic physicians, both male and female, knew of the value of the willow in pain relief. This was known across the ancient world. The treatments were complex and involved many perennial herbs, which were gathered from forest soils and from certain woodland situations that no longer exist. Most of those complex remedies have been lost. However, the use of the willow species for pain relief is widespread.

Peoples of the boreal forest system across the crown of the north still make good use of their various species of willow, as do peoples living at high altitudes in the western Himalayas. Many of these Indigenous nations have taken this knowledge underground to protect their final treasures of Indigenous ways. Finding reliable and non-toxic relief from willows for the pain of rheumatism, arthritis and osteoarthritis is still important to older people today.

The willow, all three hundred species worldwide, has a family of related medicines called salicylates. There are around a dozen biochemical components. They consist of acids, alcohols and esters. All are readily released into and around the body for absorption, which is the basis for a unique form of forest bathing practised by some Indigenous peoples of North America as a treatment for loneliness and associated mild depression. The patient is seated in a willow grove, preferably near moving water. During the day the chemicals released by the willows are absorbed through the patient's skin and lungs, and then they move throughout the body.

The Celts took their willow tree, *crann sailí*, and placed it on the podium of being sacred to their civilization. It was given the letter *S* in the Ogham alphabet and was called *sail*, represented by a vertical line with four horizontal parallel lines to the right.

Saileán was the name given to the willow bed gardens growing by their rivers and lakes. The Brehon Laws considered these gardens part of the commonage for the use of the Celts. When the willow beds were harvested by one farmer, a willow basket or large pannier would appear at the neighbour's door or half-door as an offering of appreciation. Such rites of barter and exchange were expected and when a basket arrived unannounced, usually filled to the brim with either summer butter or bastible bread, it was placed on the kitchen table and admired for at least a day.

T
Holly
Tinne

The Celts nailed their colours of red and green to the mast of their civilization. The green was not the pastoral shade of their little fields, but the proud, haughty evergreen banner of their most precious, sacred forests. The red was the scarlet of fresh blood on the battlefield or flowing from the womb at menstruation and childbirth—the liquid line between life and death, umbilical in nature, and tied to the last child born.

As a Celt, I met the holly tree. The east bedroom of the farmhouse in Lisheens was mine during the Christmas season. My mattress was one of fresh barley straw, which felt like a cloud covered with white linen sheets. It was my squeaky heaven. As the darkness of night descended, washing my window with stars, the song began from the nightingale that perched in the holly tree just outside. The river of music would be gone in the morning when the holly tree returned to sun-saturated silence.

My holly was an ancient tree and I watched it closely. As the sun climbed in the spring, the holly trees noticed; the female tree dressed itself in fragrant white flowers while the males glowed with pollen. A kind summer brought forth bunches of berries hidden beneath the evergreen leaves. As the sun backed down the horizon to shorten the days, the holly presented its ripe red berries for the feast of *nollaig*, the winter solstice, borrowed by the Christian world for Christmas.

The holly tree that was sacred to the Celts was only one of four hundred species worldwide. Theirs was the common holly, *Ilex aquifolium*, which enjoys a surprisingly large range across Europe into China. Other species, like those of North and South America, lose their leaves during the winter months, becoming deciduous trees. Wherever the holly exists in the natural world it has been sought out for its medicines by past cultures. Some of these traditional uses cling on, like *maté*, the favourite drink of South America, which is also known as Jesuits' tea.

The Druidic physicians were well aware of the healing holly, or holy tree, as it was once called. They used a tisane of the mature leaves with intact terminal spines as a tonic. It also functioned as a very mild diuretic. The tisane was used to edge down the high fevers encountered in bronchitis, and it was also used in the treatment of rheumatism.

In recent years analytical laboratories have given up the secrets of the holly. The tonic function seems to come from the plant's ability to protect the integrity of the capillaries of the vascular system. The tonic improves the movement and expansion of the capillaries as they work, feeding and oxygenating the body. Another astonishing medicine was extracted from the mature wood of the holly. This medicine was manufactured by resident fungi within the air spaces of the wood's plumbing. The endogenous fungi excreted the compound to keep the tree healthy, and the tree now offers the medicine up as a new cancer therapy currently being explored.

The Druids called their sacred holly *tinne*, representing the letter *T* in the Ogham alphabet. It was denoted by a vertical line joined by three equal parallel lines to the left. The letter was slashed into birchbark wands and then into stones, stones that have survived the assaults of time.

A Celtic Christmas features a tradition using holly boughs that has lasted thousands of years. In mid-December branches of female holly trees, selected for even sprays of berries, are gathered and brought indoors for decoration over the fire. Ivy is sometimes hung with them also. By the time the season comes to a close on January 6, with Little Christmas (also called Irish Christmas), the holly boughs have become dry. Throughout the season the decoration has released its health-giving aerosols into the atmosphere of the home. No one has been any the wiser for this working plant. Except, perhaps, the holly.

U
Heather
Úr

Heath land is one last memory held in the soil of the ancient forest system of the planet. The plant called heather, *fraoch*, with its carillon of tiny open bells, grows on the heath in the luxury of sweeping air and well-drained soil. This soil, another clue to times past, is a mixture of sharp sand and peat. The sand came from the ice-age scouring of the rocks and the peaty humus came from the mercy of the trees.

The Celts understood this landscape, filled with birds on the wing and game birds nearer the soil. Butterflies and insects abounded in this ocean of flowers that stretched into and confused the horizon. Long ago, the Druids pointed to the heath and gave it a name, *úr*, which means anything that is fresh, green and renewing. The word also came to mean free, liberal and generous. The heath, *úr*, is a place of giving for everyone, from animal to human.

The botanical body of the heath consists of a mat of very unusual plants called heathers that love a soil with the acidity of lemon juice. The plants hug the ground and carry the dark green of forest foliage. Their blooming begins as the days grow longer. The flowers, in purples and pinks edging into misty mauve, overtake the heath in a gush of wonder. They are all bells. The ling, *Calluna vulgaris*, comes first followed by the bell heathers, *Erica cinerea*, and *Erica tetralix*, the cross-leaved heather. And many more heathers come out to play on the heaths of the world.

Sometimes a mixed marriage takes place in the heather family: one set of genes changes in the course of connubial attraction and a white flower is born. These alba forms are almost as rare as hen's teeth. While all heathers have been considered to be good luck, if the white heather is given as a gift to another person, that person is considered to receive the gift of luck, a possibility of transformation for the course of their life.

I saw white heather just once. I was browsing in an antique store in the little Ontario town of Spencerville when I caught sight of a well-worn book with a faded leather cover that was scratched and torn. I opened the book and it turned out to be a Gaelic Bible written in the old style. The family and its generations of children and grandchildren adorned the title page together with a large spray of white heather, dried and pressed into the fly-leaf. It could only have come from the highlands of Scotland.

The diaspora of Scots and Irish knew the meaning behind such a sprig of heather. In the old country of their birth, it was their weather vane. On days that were warm and sunny, a mist rose from the heath, coming from the heathers. This mist veiled the mountains and the hills so that, to the human eye, it appeared as if the hills were receding and far away. This was always read as a sign of good weather in the making. If, on the other hand, the hills looked closer and sharper, rain and poor weather could be expected. The heather weather vane was used throughout the Celtic world for thousands of years by farmers and fishermen. Now we know that the mist of the heather is an aerosol vapour of arbutin and methylarbutin.

Druidic physicians understood the medical benefits of the heath. They prescribed long treks on the heath, when the heather was in full bloom, to take the air as the final healing stage for diseases of the lungs and to help redeem breathing elasticity after flus and bronchitis. Walking crushes the cuticular layer of the foliage, releasing healing aerosols. These are folded into the atmosphere and are then inhaled on the breath. The aerosol mixtures line the lungs, helping them heal. It is a form of atmospheric bathing that benefits the body.

Modern medical biochemistry comes into play to supply answers about the heath. The bell flowers of the *Ericaceae* family, including the arbutus trees of North America, Europe and Asia, produce a nectar at the base of the ovary. This sugar-rich nectar also carries the chemical calling card of arbutin. It is a foundational chemical, an antibiotic, for many of the traditional medicines of North America and elsewhere. The methylated form of arbutin, a more labile biochemical, is also produced. Like many other active drugs, arbutin can be toxic in large amounts, but the dilution factor that nature supplies in the atmosphere creates the right ratio for healing.

In addition, Druidic physicians used dark red heather honey for medicine, collected from hives whose honeybees worked the heath. In the field the bumblebees visit the heather bells first. They attack the outside petals, boring a tiny hole near the base of the ovary. Then the honeybees take over, using the convenience of the borehole to pump out the goodness. They carry it back to the hive where the bees store it in separate combs and reduce it to a dense honey that has an extremely high viscosity. It is almost a solid that contains the totality of resins also collected on the heath. Heath honey was used on its own as a medicine for sore throats and colds.

Sometime in the distant past, the Druids pointed to their sacred landscape, the heath, *úr*. This was their field of freedom, *saoirse*, so precious to the Celtic culture. They used *úr* for the letter *U* in the Ogham script. It carried all that was important to the Celtic heart.

Z
Blackthorn
Straif

The man who walked with the dragon stick always came speaking of marriage. He approached the farming settlement slowly, as if he were having some ponderous thoughts. His eyes were everywhere, looking for quality, conduct, sleek coats on the calves, the degree of hospitality laid down by Brehon Laws. His gaze ran over the fields and experience spoke to him of the quality, or lack of it, of the land. Land was everything in the Celtic world.

The man was the matchmaker, the *babhdóir*. The first matchmaker who came to the farmhouse in Lisheens when I was a child took me by total surprise. He looked like a big, blond bear but he carried a traditional shillelagh in his hand. I was sent to sit in the parlor. I heard arguments. I left the parlor and crept to the kitchen door where I heard my pedigree being recited like a poem. My aunt Nellie told the *babhdóir* in no uncertain terms that marriage was not yet on the table for me. I had not finished my Brehon wardship. I was to be educated, and that education was to go as far as was possible. I never saw that blond *babhdóir* again. Nobody would ever speak of that sunny afternoon when loose, black tea was served in china teacups, on old Irish linen, to a man who carried a shillelagh.

But of course, as boys and girls grew up, marriage was on their minds. Sometimes a wedding came in the slipstream of love and other times not. It always came together by arrangement, though, because family connections were important. Both parties had to agree. There was considerable debate in the community about character qualities of the groom and bride-to-be, which were taken into consideration by both sets of parents. Meetings were arranged and a dowry price discussed.

The agreement of the dowry was prenuptial and always in the bargaining hands of the *babhdóir*. The goods and services of the dowry was the value that the bride and groom had in their own home. Together and in combination, this was land, cattle, money or other goods. A woman could ask for a man's hand in the union of marriage with equal flair to a man.

The matchmaker, *babhdóir*, was always a man of good standing in the community. He could be trusted to be silent if necessary and not to release any family skeletons into the general public during the course of his business. As he was invited into the kitchen, he placed the dragon stick across the table so that the household understood the intent of the visit. If the stick was gently and carefully ignored, then there were other offers in the making for marriage and he had to wait his turn to be told of these. Sometimes the parties picked up the stick and placed it against the wall, showing him without saying it that there was interest in the subject of marriage and the visit was simply enjoyed by everyone present, as was the custom.

Marriage for the Celts was a contract that had much to do with the protection of land, which was the lifeblood of the community. Marriage alliances were taken seriously because there were obligations of kinship, *gaol*, attached to both families. Marriages bonded the community, knitting a family pattern of personal support within the Brehon laws of hospitality. And of course divorce was always possible, if it became necessary.

The matchmaker also had the delicate matter of breeding lines, *pór*, to consider. The Celts were obsessed with breeding, be it of seeds, apples, cattle, dogs or horses. They put as much care into the matching of family traits as they could. Intelligence, diligence, work, smithing, memory, oratory and the mathematics of mind riddles were well placed in the appropriate family lines. Much care was taken to maintain and protect the leading lights of the civilization, especially in literature, poetry, music and the arts.

The dragon stick was cut from a magical tree commonly called the sloe in Ireland. This small, spiny tree is also called the blackthorn or *Prunus spinosa*. It is the source of traditional blackthorn sticks, which are also called Irish shillelaghs, directly from the old Irish, *sailéille*, meaning a cudgel. These sticks were used by cattle drovers as well as matchmakers, and also in the king's court as an instrument to beat time to music. The mixture of its deep tone mixed with the dry rattle of the bones and the winnowing drum, the *bodhrán*, are still a signature of authentic Irish music.

The blackthorn trees must have calcium in the soil to produce a crop of sloe plums, which hang on the trees until November when the cool nights and early frosts change the sugar content of the small black fruit—a walker's trail food picked from the hedgerows. The surface yeasts multiply also, and the wine produced by fermentation was passed around from house to house and enjoyed by all. The fruits also produced a sloe gin.

The little scattered blackthorn was designated a sacred, magical tree and placed in the Ogham script, representing the letter *Z*, and called *straif*, for its straggling and generally untidy appearance. It is depicted as a vertical line intercepted by four parallel lines that tilt up to the left.

The Druids who practised magic loved this tree. They used the dragon stick or shillelagh as a form of a wand of authority when they practised their art.

ACKNOWLEDGMENTS

I WANT TO THANK the host of people in my life who made this book possible. In the Valley of Lisheens in County Cork all of them have gone. But they have left behind a great warmth in my heart for what they gave me.

My main editor, Evan Rosser, somehow managed to listen to the many traumas of my early life and make some sense out of it all. I am grateful for that. Anne Collins, my publisher at Random House Canada, maintained an editorial tapestry of words that stitched science into the sentences. Stuart Bernstein, my agent, is always on tap for advice, conversation and protection—remember that policeman?

Lynn and Nancy Wortman provided tea and typing in a haven of laughter. Tilman Lewis is an impressive copy editor; Lisa Jager, designer of the Random House Canada edition, has my deep gratitude.

I want to also thank Tom Fischer, Alex Fus, Adrianna Sutton, and Michael Dempsey of Timber Press for helping to spread a bigger message of climate change across the world.

My children, Erika and Terry, inspire me with their love and confidence. I want to especially thank my husband, Christian H. Kroeger, for support that never fades.

A READING LIST

Barnhart, Robert K. *Chambers Dictionary of Etymology*. London: Chambers Harrap, 1988.

Beresford-Kroeger, Diana. *Arboretum America: A Philosophy of the Forest*. Ann Arbor: University of Michigan Press, 2003.

Beresford-Kroeger, Diana. *Arboretum Borealis: A Lifeline of the Planet*. Ann Arbor: University of Michigan Press, 2010.

Beresford-Kroeger, Diana. *A Garden for Life: The Natural Approach to Designing, Planning, and Maintaining a North Temperate Garden*. Ann Arbor: University of Michigan Press, 2004.

Beresford-Kroeger, Diana. *The Global Forest*. New York: Viking, 2010.

Beresford-Kroeger, Diana. *The Medicine of Trees: The 9th Haig-Brown Memorial Lecture*. Campbell River, British Columbia: Campbell River Community Arts Council, 2018

Beresford-Kroeger, Diana. *The Sweetness of a Simple Life*. Toronto: Random House Canada, 2013.

Chadwick, Nora. *The Celts*. London: Folio Society, 2001.

Conover, Emily. "New Steps Forward: Quantum Internet Researchers Make Advances in Teleportation and Memory." *Science News,* October 15, 2016, 13.

Conover, Emily. "Emmy Noether's Vision." *Science News*, June 23, 2018, 20-25.

Cross, Eric. *The Tailor and Ansty*. 2nd ed. Cork: Mercier Press, 1964.

Daley, Mary Dowling. *Irish Laws*. San Francisco: Chronicle Books, 1989.

De Bhaldraithe, Tomás. *English-Irish Dictionary*. Dublin: Cahill, 1976.

Ellis, Peter Berresford. *A Brief History of the Celts*. London: Constable and Robinson, 2003.

Fulbright, Dennis W., ed. *A Guide to Nut Tree Culture in North America*. Vol. 1. East Lansing: Northern Nut Growers Association, 2003.

Ginnell, Laurence. *The Brehon Laws: A Legal Handbook*. Milton-Keynes: Lightning Source UK, 2010.

Hamers, Laurel. "Quantum Data Locking Demonstrated: Long Encrypted Message Can Be Sent with Short Decoding Key." *Science News*, September 17, 2016, 14.

Herity, Michael, and George Egan. *Ireland in Prehistory*. London: Routledge, 1996.

Hillier, Harold. *The Hillier Manual of Trees and Shrubs*. Newtown Abbot, UK: David and Charles Redwood, 1992.

"Hydrological Jurisprudence: Try Me a River." *The Economist*, March 25, 2017, 34.

Jacobson, Roni. "Mother Tongue: Genetic Evidence Fuels Debate over a Root Language's Origin." *Scientific American*, March, 2018, 12–14.

Kotte, D., Li, Q, Shin, W. S. and Michalsen, A. (eds.). *International Handbook of Forest Therapy*. Newcastle upon Tyne, UK; Cambridge Scholars Publishing, 2019

Lewis, Walter H. and P. F. Elvin-Lewis. *Medical Botany: Plants affecting Man's Health*. 2nd ed. Toronto: John Wiley and Sons, 2003.

Liberty Hyde Bailey Hortorium. *Hortus Third: A Concise Dictionary of Plants Cultivated in the United States and Canada.* New York: Macmillan, 1976.

Ó Dónaill, Niall. *Foclóir Gaeilge-Béarla.* Dublin: Richview Browne and Nolan, 1977.

O'Neil, Maryadele J. *The Merck Index: An Encyclopedia of Chemicals, Drugs, and Biologicals.* 14th ed. Whitehouse Station, NJ: Merck, 2006.

Stuart, Malcolm. *The Encyclopedia of Herbs and Herbalism.* London: Orbix, 1979.

Tree Council of Ireland. *The Ogham Alphabet.* Enfo: Information on the environment, undated.

INDEX

Dr. Diana Beresford-Kroeger is a world-recognized botanist, medical biochemist and author, whose work uniquely combines western scientific knowledge and the traditional concepts of the ancient world. Her books include *The Sweetness of a Simple Life*, *The Global Forest*, *Arboretum Borealis*, *Arboretum America*—which won the National Arbor Day Foundation Award for exemplary educational work on trees and forests—*Time Will Tell* and *A Garden for Life*. Among many honours, Beresford-Kroeger was inducted as a Wings World Quest fellow in 2010 and elected as a fellow of the Royal Canadian Geographical Society in 2011. More recently, in 2016, the Society named her one of 25 women explorers of Canada. Her work has inspired artists and writers, as well as other leading scientists. She is the author and presenter of a feature documentary, *Call of the Forest*, and is also at the heart of the PBS series, *The Truth about Trees*. Currently she is advocating on behalf of an ambitious global "bioplan" encouraging ordinary people to develop a new relationship with nature and join together to restore the global forest.